ASIAPAC CULTURE

Origins of CHINESE SCIENCE and TECHNOLOGY

Compiled by Asiapac Editorial
Illustrated by Fu Chunjiang Translated by Yang Liping and Y N Han

ASIAPAC • SINGAPORE

Publisher
ASIAPAC BOOKS PTE LTD
996 Bendemeer Road #06-09
Singapore 339944
Tel: (65) 6392 8455
Fax: (65) 6392 6455
Email: asiapacbooks@pacific.net.sg

Come visit us at our Internet home page
www.asiapacbooks.com

First published July 2004
4th edition January 2008

ISBN 13 978-981-229-376-3
ISBN 10 981-229-376-0

Cover illustrations by Fu Chunjiang
Cover design by Kelly Lim
Body text in 11pt Times New Roman
Printed in Singapore by FuIsland Offset Printing (S) Pte Ltd

Publisher's Note

China is the oldest continuous major civilisation in the world, traditionally held to be 5,000 years old, with written records dating back 3,500 years. As such, it is best known for its cultural and artistic achievements.

Lesser known are the Chinese people's countless ingenious inventions and scientific discoveries, with great minds making major contributions in every field from mathematics to astronomy to geography and agriculture far ahead of their time. Yet, modern science and technology is typically associated today with the western world, by the Chinese as well, even as scientists rediscover the knowledge handed down from the distant past and validate long-held beliefs in the laboratory.

With *Origins of Chinese Science and Technology*, we hope to bring to light the fascinating legends and often surprisingly modern ideas about science and technology developed by the Chinese people. Each chapter covers a different field of knowledge, clearly explained with lively and easy-to-understand illustrations.

This book is the product of a collaborative effort on the part of the Asiapac editorial team to research and compile materials. We wish to thank Mr Fu Chunjiang for his vibrant illustrations, and Mr Yang Liping and Ms Y N Han for the translation. We also would like to thank Ms Grace Chiam and Ms Sin Yoke Yin for their editorial support. Finally, we wish to thank the production team for their hard work that has made this publication possible.

CHINESE CULTURE SERIES

150x210mm, fully illustrated, 160-192 pages

978-981-229-475-3	Origins of Chinese Music
978-981-229-462-3	Origins of Chinese Names
078-981-229-242-1	Origins of Chinese People & Customs
978-981-229-378-7	Origins of Chinese Festivals
978-981-229-407-4	Origins of Chinese Culture
978-981-229-318-3	Origins of Chinese Food Culture
978-981-229-317-6	Origins of Chinese Cuisine
978-981-229-369-5	Origins of Chinese Tea & Wine
978-981-229-441-8	Origins of Chinese Art & Craft
978-981-229-376-3	Origins of Chinese Science & Technology
978-981-229-268-1	Origins of Chinese Martial Arts
978-981-229-408-1	Origins of Shaolin Kung Fu

About the Illustrator

Fu Chunjiang 傅春江 was born in Chongqing, China and received his BA in Chinese language and literature. A lover of Chinese culture, he is a versatile artist skilled in different styles, in particular the traditional style of comic books adapted from Chinese classics. His illustrations feature minute strokes, and clear and bright lines that can communicate the characters' rich and sincere feelings.

He has so far illustrated the following popular Asiapac titles: *Chinese Culture Pack*, *Origins of Chinese People and Customs*, *Origins of Chinese Culture*, *Gateway to Chinese Classical Literature*, *Essence of Traditional Chinese Medicine*, *Origins of Chinese Science and Technology*, *Origins of Chinese Festivals*, *Origins of Chinese Food Culture*, *Origins of Chinese Tea and Wine*, *Stories of Honour,* and more.

Contents

Prologue

China is the cradle of inventions in the world... and its science and technology had reached a level far higher than that of the West between the 3th century BC and the 13th century AD...

~ Joseph Needham (1900-1995)

The ancestors of the Chinese people had to face many threats in the course of their existence, both from nature and from wars. In order to make their way in the world, they had to rely on their ingenuity and constantly increase their store of knowledge, stimulating the development of science and technology.

The importance of science and technology to Chinese culture can be clearly seen in their ancient myths and legends, which are filled with stories of how various 'deities' or 'sages' invented all sorts of things in order to improve the lives of people. Examples include Suiren drilling wood to start fires; Fuxi domesticating beasts; Shennong tasting all types of plants to learn about food plants, medicinal plants and poisons; the many important inventions attributed to Huangdi and his court and more.

With its long agricultural tradition, China is credited with the invention of many sophisticated farming techniques and tools, abetted by its industrial expertise. It has domesticated numerous plant species and developed advanced techniques for working bronze and manufacturing steel.

Other important achievements in the fields of science and technology include the Four Great Inventions — the compass, paper-making, gunpowder and printing — which have greatly advanced the course of civilisation in the world.

Chinese astronomers have made close observation of the skies and astronomical phenomena for thousands of years, recording eclipses, sunspots and comets as early as two thousand years ago. Zhang Heng, a scientist of the Western Han Dynasty, constructed an armillary sphere and also invented the first seismograph in the world.

In mathematics, China was the first to adopt the decimal system. It also invented the abacus, a simple calculation tool that is still highly valued today. Zu Chongzhi, a mathematician living during the Southern and Northern Dynasties, achieved a determination of *pi* accurate to six decimal places, ahead of the world by up to 1,000 years.

Great building projects completed by China include the Great Wall of China, Dujiang Dam and the Grand Canal that runs through Beijing and Hangzhou, the longest and oldest canal in the world. All these show the high technological levels attained by the early Chinese.

The rich and colourful Chinese science and technology has spread through the world via many channels such as the Silk Road and contributed considerably to the advancement of human civilisation.

FOUR GREAT INVENTIONS

China boasts four great inventions: the compass, paper-making, gunpowder and printing. These inventions have spread from China to other parts of the world and made considerable contributions to the progress of civilisation.

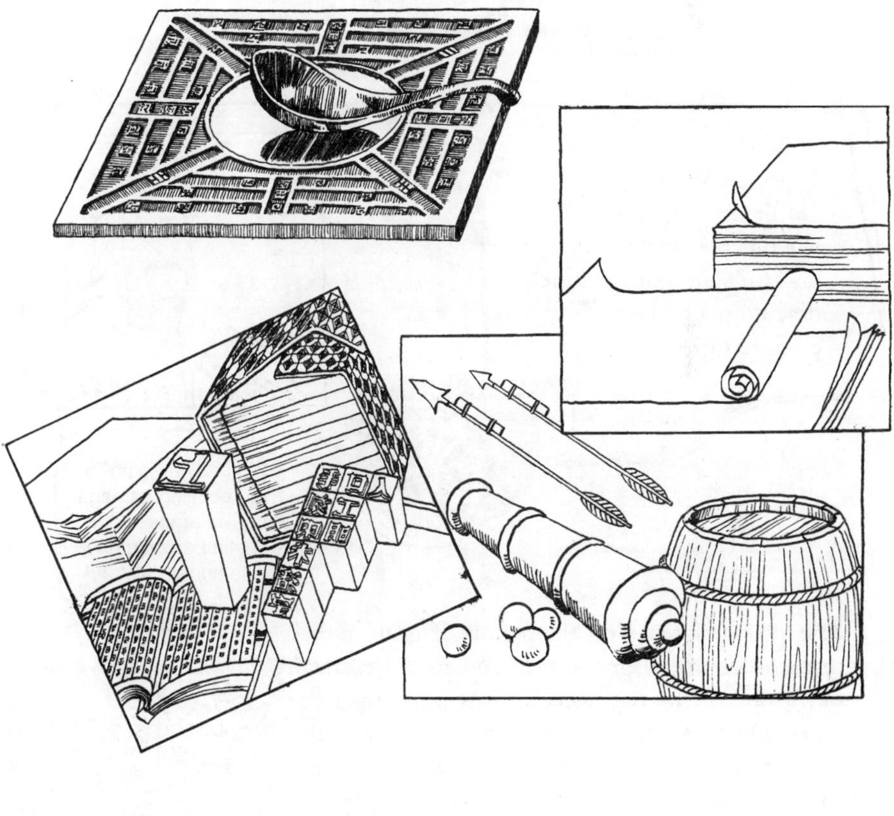

Compass

How the South-pointing Carriage Works

On the carriage stands a wooden person pointing his arm southwards all the time no matter how the carriage turns.

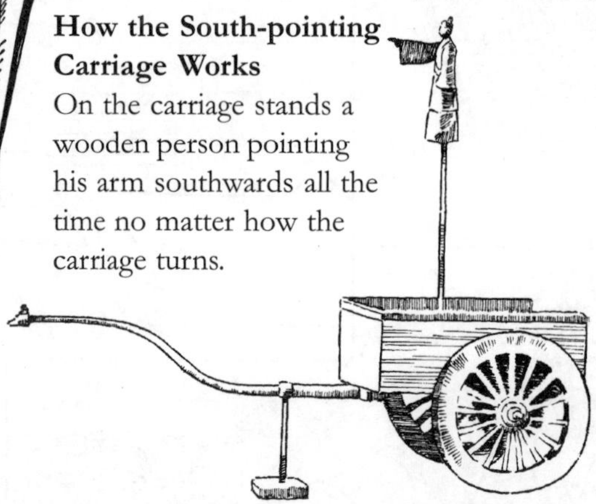

This property relies on the action of gears rather than the earth's magnetic field. When the carriage turns, one of the two wheels turns more than the other, which causes the gears to turn the arm in the opposite direction, ensuring that it always points southwards.

Magnetic Compasses

The first compasses were made from naturally occuring lodestone, which is a kind of iron ore with magnetic properties. Over time, people also found other ways they could make their own magnets from iron.

Si Nan or South-Pointing Ladle

Invented about 2,000 years ago, *si nan* is the earliest compass made of lodestone. The lodestone was shaped into a ladle and placed on a smooth copper plate with 24 direction marks on it. When spun, it always comes to a halt pointing to the south.

South-Pointing Fish

During the Song Dynasty, people invented an artificial magnet called south-pointing fish.

How a South-Pointing fish is made

1. Take a piece of fish-shaped iron sheet.

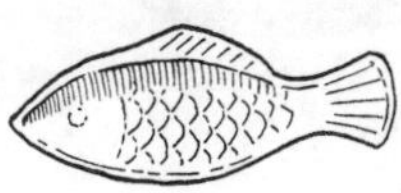

2. Heat it in a fire until it is red-hot.

3. Place it on the ground with its head pointing south and its tail pointing north.

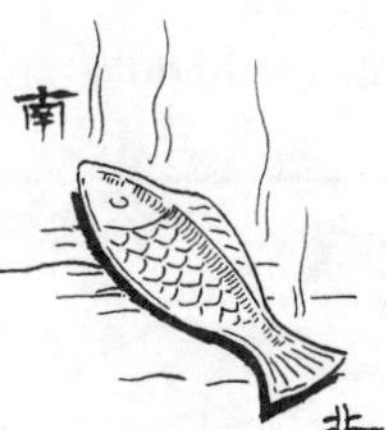

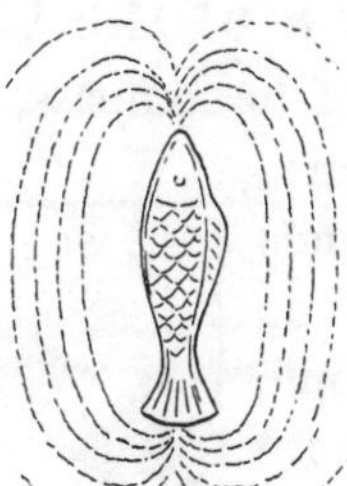

4. The iron will be magnetised by the earth's magnetic field.

South-Pointing Turtle

Inside the turtle's stomach is a lodestone. A hole is made in its underbelly so that it can be balanced on a bamboo nail. The turtle can spin freely and will point to the south when it comes to a stop.

Plate Compass

We can also magnetise a needle by stroking it with a piece of lodestone. During the Southern Song Dynasty, the magnetic needle was combined with a plate with divisions for measuring angles. This was the prototype for the modern compass.

The Contribution of the Compass

Before the invention of the compass, sailors could only navigate by the positions of the sun and stars, which were sometimes blocked by clouds. The compass allowed people to draw more accurate charts and travel freely on open sea, opening the way to exploring the world and making international contacts. It reached the Arab world in the 12th century, and Europe in the 13th century.

Magnetism

Each magnet has two poles — the North (N) Pole and the South (S) Pole. ▽

Two like poles repulse each other... ▽

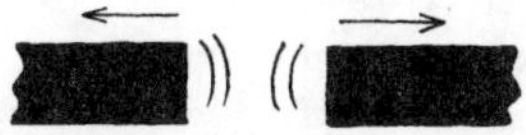

...while two unlike poles attract each other. ▽

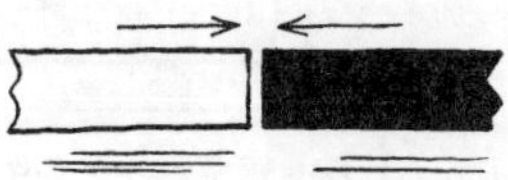

△ The earth has a natural magnetic field with two poles. Due to this magnetic field, magnets always point in the same direction.

A magnet always has a north-seeking pole and a south-seeking pole. Thus, it can be used to tell the direction.

Guarding Against Assassins with Magnets
Following his unification of China, Emperor Shihuangdi of the Qin Dynasty built the grand E-Fang Palace. To prevent assassins, the palace gates were made of magnets which could attract and detect all sharp weapons hidden by the would-be assassins. This shows that the Chinese people had already recognised the basic properties of magnets as far back as 2,000 years ago.

Zheng He's Voyages to the West

The renowned Ming Dynasty navigator Zheng He used the compass during his voyages to the West.

Zheng He was born to the Hui ethnic group. He was also known as "Eunuch Sanbao" as he was a eunuch in the imperial palace.

To promote foreign trade, the Ming emperor sent him on missions to the area west of the South China Sea, including the Indian Ocean and its coastal regions. In July 1405, Zheng He embarked on his first voyage at the head of the biggest fleet in the world at that time, with more than 200 ships.

Not only could the compass that the fleet used tell the direction, it could also measure the distance travelled by the ships.

During the seven voyages made over 28 years, Zheng He's fleet left its mark in the Indochinese Peninsula, the Malay Peninsula, the Nansha Islands, India, Persia, the Arabian area, the Eastern African coast and the Red Sea. It had visited more than 30 countries and regions, and expanded China's trade with Southeast Asia.

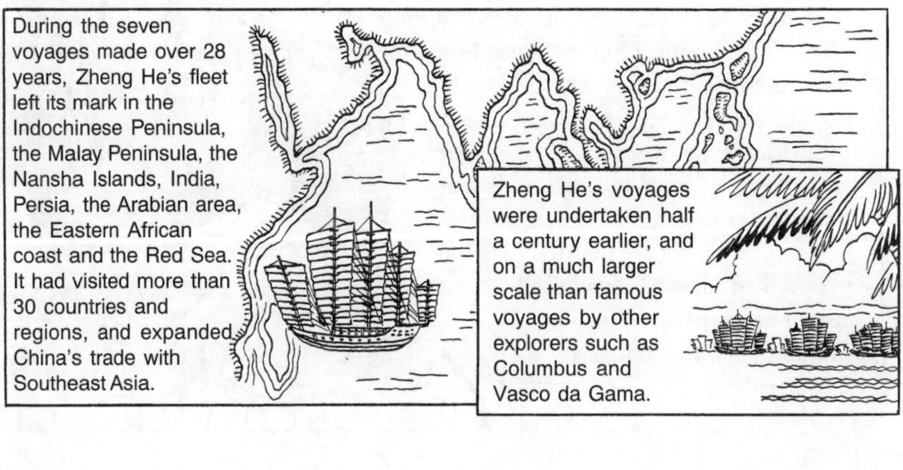

Zheng He's voyages were undertaken half a century earlier, and on a much larger scale than famous voyages by other explorers such as Columbus and Vasco da Gama.

Paper-Making

Before Paper

Before paper was invented, people used knives and drills to inscribe their words on oracle bones and bronzeware.

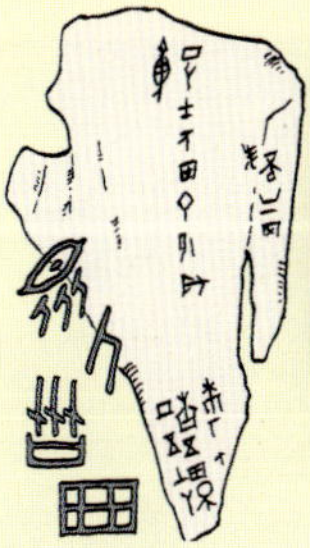

Later, people started to write on bamboo or wooden slips. These slips could be tied together into a scroll.

These scrolls were heavy and bulky, and quite inconvenient to carry around. Emperor Qin Shihuang was said to have read 120 kg of official documents every day! During the Western Han Dynasty, Dongfang Shuo once wrote a 3,000-slip letter to the emperor, which had to be carried to the imperial palace by two warriors and took the emperor two months to read.

During the Qin Dynasty, people began writing on fine silk instead. Though it was convenient to write on and easy to transport, silk was too expensive for ordinary people.

Five Cartloads of Learning

Hui Shi was a well-known scholar of the Warring States Period. He would take his students and books with him wherever he went. One cart was enough for him and his students, but he needed five carts for his books, which consisted of bamboo slips. Later, people used the phrase 'five cartloads of learning' to describe a learned man who has read widely.

The Earliest Paper

Paper appeared during the Western Han Period. Made from plant fibers, the earliest paper was rough in quality and difficult to write on. In 1957, archaeologists uncovered 88 pieces of ancient hemp paper from an ancient tomb in Baqiao, east of Xi'an. This is the earliest known paper.

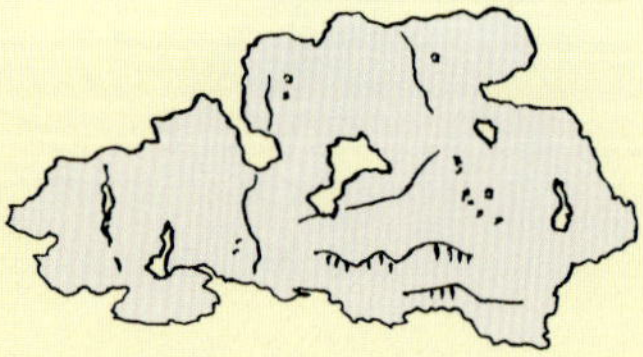

Cai Lun Makes Paper

In AD 105, Cai Lun, a confidential secretary to the emperor during the Eastern Han Period, studied the experiences of his predecessors and revolutionised the paper-making technique to come up with a high quality paper made from bark, rough hemp, rags and old fishnets. Thin, light and strong, the paper was easier to write on and much cheaper. It soon became popular. Later, it was called "Marquis Cai's Paper".

Cai Lun's Paper-Making Process

1. First, soak hemp, rags and other materials in water.

2. Cut them into pieces and cleanse with water.

3. Immerse the pieces in a mixture of water, grass and tree ash, and then stew them.

4. Rinse the mixture with clean water and pound it to obtain thin fibres.

5. Fish out the fibres with a paper matrix.

6. Dry them in the open and the result is paper.

The Spread of Paper

Paper was not widely used before Cai Lun's innovation. Heavy bamboo slips and expensive silk did not give way to paper until his new paper-making technique allowed high quality paper to be produced economically.

China's paper-making technique spread to Korea and Vietnam in the 4th century, to Japan and the Arabian area in the 7th century, and finally to Europe and Africa in the 12th century, thus speeding up the advancement of civilisation.

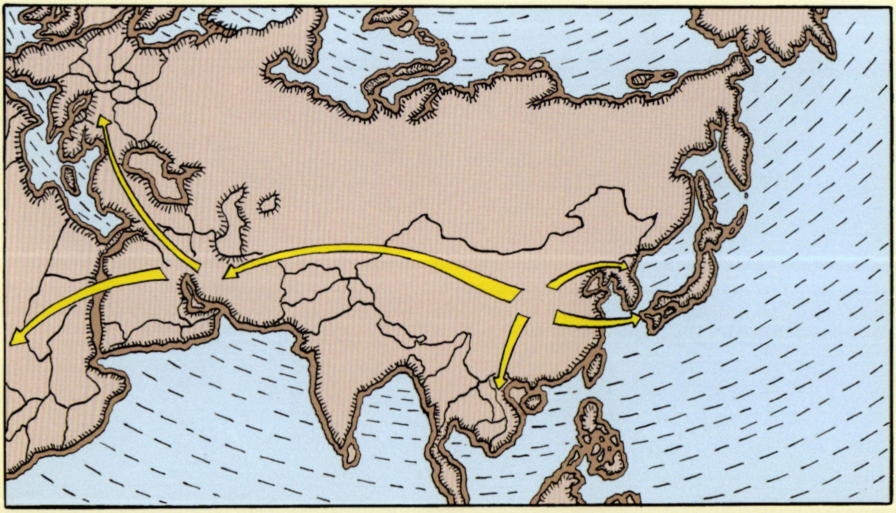

How did other countries make written records before the arrival of China's paper-making technique?
India — birch bark and palm leaves.
Egypt — papyrus.
Made from a plant found in the valley of the Nile river, the paper is heavy and fragile.
Europe — animal skin parchment. Transcribing the Holy Bible took more than 300 pieces of lambskin. Ordinary people could not afford this.

Modern Paper-Making

Gunpowder

Gunpowder is made from saltpetre, sulphur and charcoal. *Classic of Herbal Medicine* (*Shennong Bencaojing*), a Han Dynasty medical classic, lists sulphur and saltpetre as important medicines. The Ming Dynasty physician Li Shizhen believed that gunpowder could cure ringworms, kill pests, and even prevent humidity and pestilence.

The Discovery of Gunpowder

The process of making gunpowder was discovered by ancient Taoist priests who wanted to make pills of immortality. They mixed sulphur and saltpetre with other materials and heated the mixture in a stove. Instead of getting pills of immortality, they discovered an explosive! So gunpowder was produced accidentally.

Use of Gunpowder

Gunpowder was invented over 1,000 years ago when the Sui Dynasty gave way to the Tang Dynasty. Towards the end of the Tang Dynasty, gunpowder-based weapons such as rockets were already in use. Besides military applications, gunpowder was also used widely in entertainment and engineering, finding its way into fireworks, mining, railway-building, tunnel-digging and water control projects.

Between the 12th and the 13th centuries, the use of gunpowder first spread to the Arabian countries, then to Greece and Europe.

Taoist Alchemy

In addition to their accidental discovery of gunpowder, Taoist priests also accumulated tremendous knowledge of chemistry in their attempts to create pills of immortality. Alchemy spread to Europe via the Arabs in the 8th century and the Europeans transformed it into pharmaceutical chemistry.

Weapons Using Gunpowder

Gunpowder allowed for the development of many powerful weapons.

Firelocks

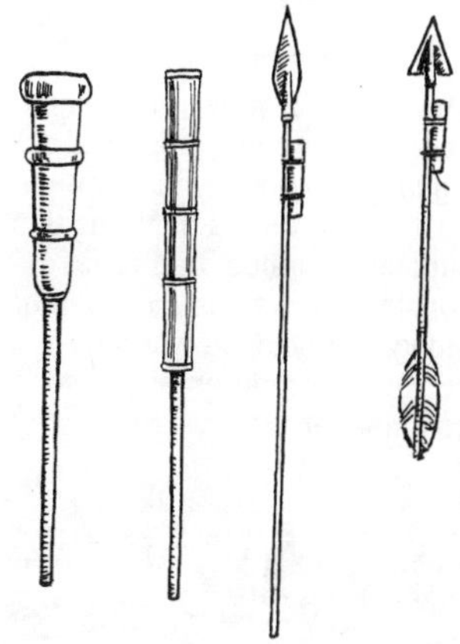

The Chinese invented the firelock in AD 905. It was a spear with a powder flask tied to it. A soldier could ignite the powder while charging towards the enemy. During the Southern Song Period, firelocks made with bamboo poles were used. By the Yuan Dynasty, metal tubes had replaced bamboo ones. Hence bronze and iron firelocks appeared, very much resembling today's rifles.

Cannons

A legend has it that an ancient farmer invented a type of 'stone flinger' that was capable of throwing stones more than 250 metres away. Later, strategists used the device to hurl explosive satchels. This was the beginning of cannons.

A highly skilled artisan named Ma Jun built a kind of wheeled catapult that could fling as many as 72 explosive satchels in quick succession.

The cannon was invented during the Yuan Dynasty. It could be mounted on a camion and fired at targets thousands of metres away. Later, foreign missions brought this weapon to Europe, where European artisans invented the linear-bore gun, which could stabilise the path of the cannonball.

Catapult

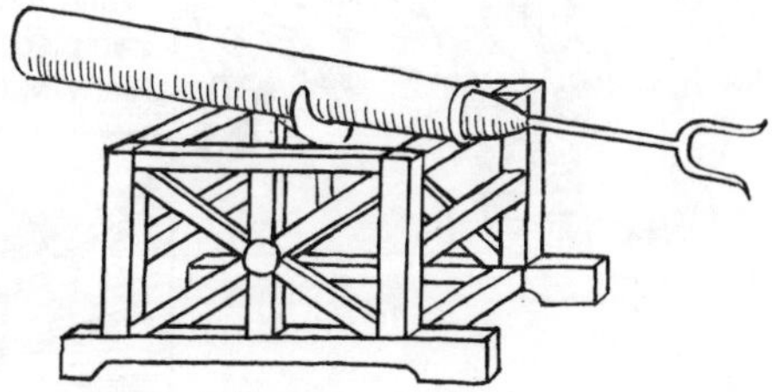

Cannon

Invention of the Mine

The Chinese were using mines 300 years before the Europeans did. Qi Jiguang, a Ming Dynasty general best known for his battle against Japanese pirates, had to face constant incursions by mounted Mongolian soldiers when he served in Jizhou as a garrison commander.

He gave orders to build special wooden boxes fitted with special steel wheels and flintstones, which would create sparks when the boxes were stepped on.

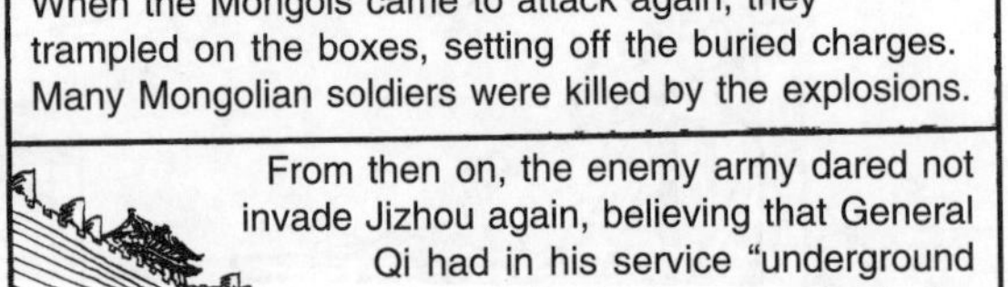

Printing

Before the Invention of the Printing Press

The oldest books in China were hand-copied before the invention of the printing press. It was both time- and energy-consuming, and spelling mistakes occurred quite often. Later, people learned to rub seals on stone tablets. This is arguably the earliest printing method.

The Process of Tablet-Rubbing

1. Sprinkle a piece of paper with clean water.

2. Cover the tablet with the paper and smooth it out.

3. Tap the paper with a bundle of cloth and the paper become dented where there are characters.

4. Dip the cloth into ink, daub the paper black and the characters will appear.

Block Printing

Surface printing was invented during the Sui and Tang Dynasties.

1. Carve reversed characters in relief on a wooden block.

2. Apply black ink onto the block.

3. Press a sheet of paper on it to make a copy.

The Diamond Sutra (*Vajracchedika-Prajna-Paramita-Sutra*) was printed during the Tang Dynasty using the woodcut technique. It is the earliest known printed work.

Invention of Movable Type

During the Northern Song Dynasty between 1041 and 1048, Bi Sheng, a carver with a book printing house in Hangzhou, invented movable type.

1. He made squares out of sticky clay and carved individual characters on them.

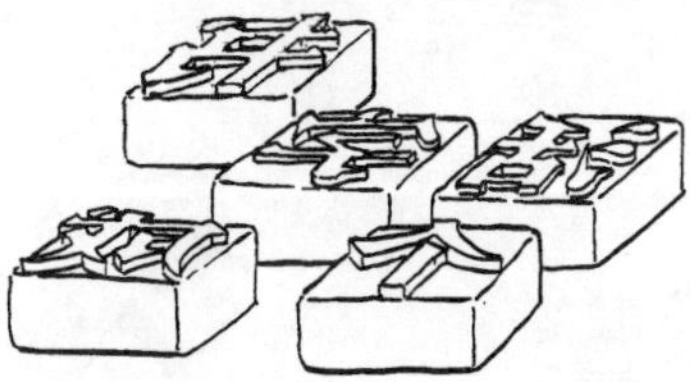

2. Then, he baked the blocks in fire to harden them.

3. The clay blocks were arranged on a framed iron board for printing.

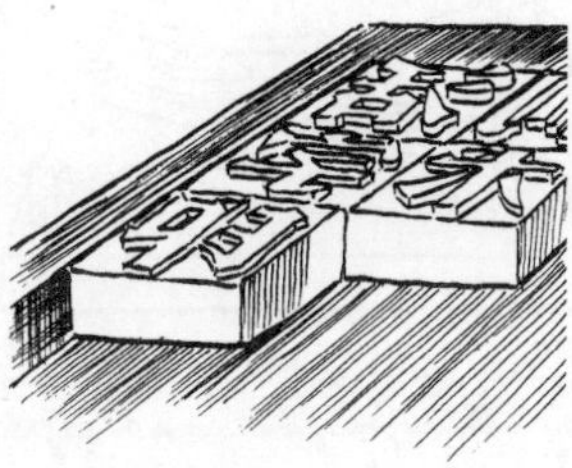

4. Pressure was applied to smoothen and secure the frame.

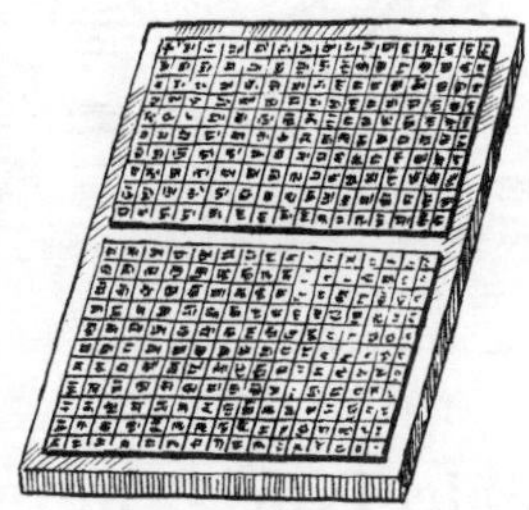

5. The characters were daubed with ink and printing could start.

6. After printing, the clay blocks could be taken out of the frame and repositioned for other pages and books. It proved to be an effective way to save time and energy since people did not have to carve new blocks for plate-making. Even if a wrong character was spotted, it could be corrected and replaced at any time. Hence, printing quality greatly improved. It was not long before movable-type printing replaced woodcut printing, opening the way for greater cultural communication.

Revolving Typesetting Plate

Wang Zhen of the Yuan Dynasty invented the revolving typesetting plate. Character types were arranged alphabetically on the revolving plate. The plate could be revolved to look for types in the course of type-setting. A worker could sit on a chair while picking out the types. It was quite convenient and efficient.

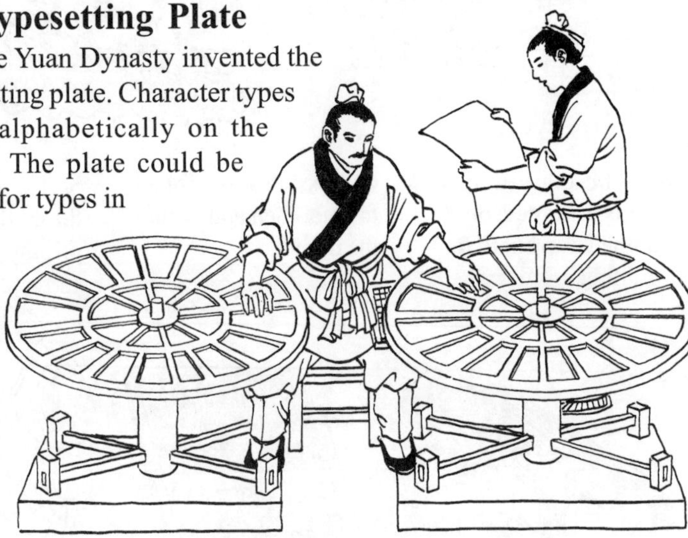

The Spread of Movable-Type Printing

Movable-type printing spread to Korea, Japan, Vietnam and the West in the 13th century. In 1450, or about 400 years after Bi Sheng's invention, Johannes Gutenberg from Germany printed *The Holy Bible*. It was the first book printed in Europe with the movable type.

ASTRONOMY

Astronomy is an ancient science in China. The early Chinese scientists invented some of the first astronomical instruments in the world. Not only were their astronomical records dated much earlier, they were more complete, detailed and reliable than those of other civilisations.

THEORIES ON THE STRUCTURE OF THE UNIVERSE

The ancient Chinese developed three theories on the structure of the universe.

The Heavenly Cover

Heaven covers the earth like a dome while the earth below is flat like a chessboard.

Infinite Empty Space

Heaven is a boundless empty space filled with air, in which the sun, moon and stars float freely.

The Celestial Sphere

Space as a whole is likened to an egg with the earth at its core, as its yoke.

Pan Gu Separates Heaven from Earth
In the beginning...
Heaven and earth were mixed up like an egg.
Pan Gu was born in between.
After 18,000 years, Pan Gu woke andbroke open the egg. The bright and light part rose to become the sky, while the dark and heavy part sank and became the earth.
To keep heaven and earth from coming together again, Pan Gu used his arms to hold up the sky while pressing down on the earth with his feet. After tens of thousands of years, they were finally completely separated.
Pan Gu died of exhaustion. After his death, his eyes became the sun and moon, his body the mountains, his blood the rivers, his hair the vegetation and his sweat the rain and mists.

Methods of Dividing the Starry Sky

Twenty-Eight Constellations

The ancient astronomers divided all the visible stars in the sky into 28 constellations, evenly distributed in the east, south, west and north. The eastern part was named the Green Dragon (*Cang Long*), the western part White Tiger (*Bai Hu*), the southern part Red Sparrow (*Zhu Que*) and the northern part the Black Turtle (*Xuan Wu*).

Three Walls

The ancient people also felt that the stars were arranged like walls in some parts of the sky. So they called them Taiwei Wall, Ziwei Wall and Tianshi Wall. The ancients divided the sky by these three walls and the 28 constellations.

The Cowherd and the Weaving Girl

As they gazed at the stars in the sky, the ancient Chinese were inspired and spun many stories. The most romantic tale tells about a cowherd and a weaving girl.

Once upon a time, a weaving girl descended from heaven to the earth. She fell in love and married a cowherd. They had two children, a boy and a girl, and were very happy together.

When the heavenly emperor's mother learnt of this, she came with a regiment of heavenly soldiers to bring the weaving girl back by force.

The cowherd put his children in two baskets and carried them with a shoulder pole as he pursued the soldiers to heaven.

The emperor's mother made a torrential river with her magical hairpin, separating the couple.

Moved by their love, magpies would appear on the seventh day of the seventh month every year and build a bridge with their bodies so that the cowherd and the weaving girl could be reunited.

The river in the story actually refers to the Milky Way, a part of which appears as a bright band of stars across the sky. To the west of the river is a bright star called Vega. Beside it are four stars which look like a shuttle if connected. Thus they are called the shuttle stars. To the east of the river is Altair, a yellow star. Near it are two little stars. They are joined to form a shoulder pole and are called the shoulder-pole stars. The Cygnus constellation, which looks like a swan, is the legendary bridge of magpies.

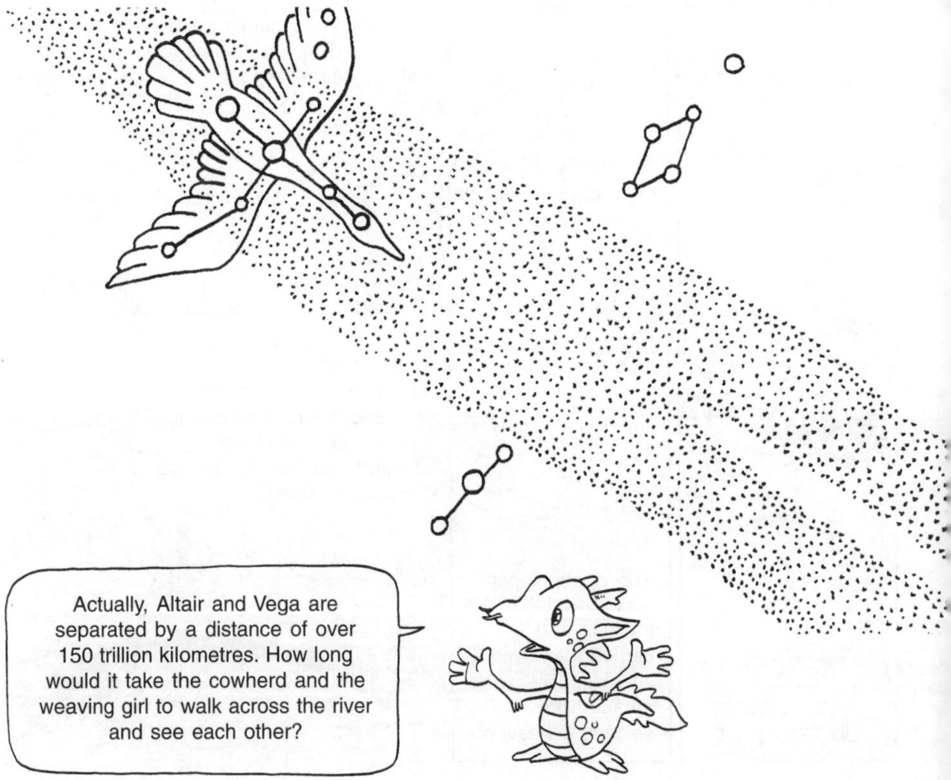

Astronomical Observations

Sunspots (Maculae)

Sunspots were recorded in Chinese documents over 2,000 years ago. The spots were observed and recorded systematically. Their changes and movements were also noted. In Europe, sunspots were not discovered until 1610 when Galileo observed the sun with the telescope he invented.

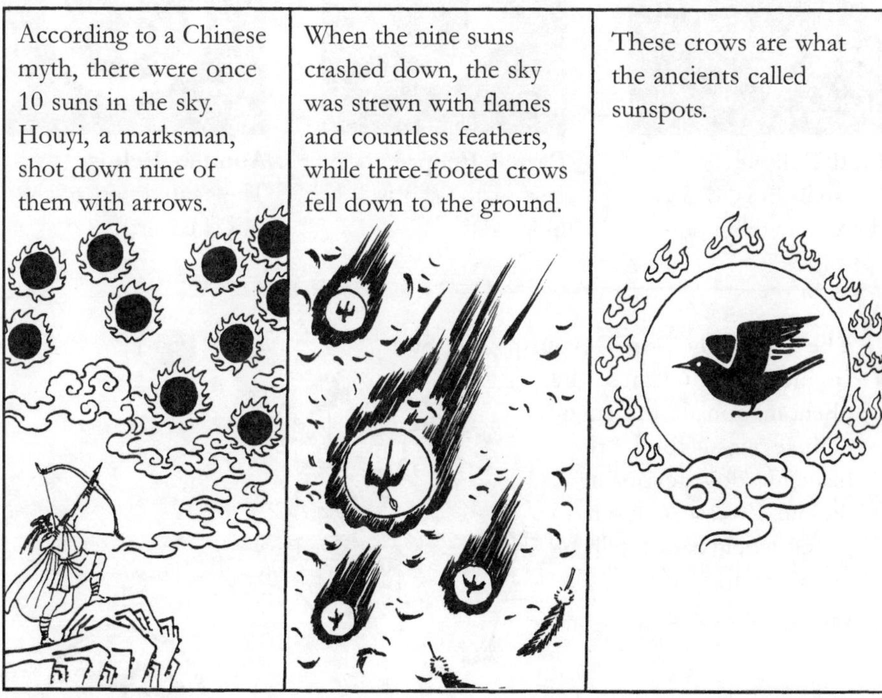

Solar Eclipses

The Chinese classic, *Shangshu*, or *The Book of History,* mentioned a solar eclipse that occurred during the Xia Dynasty. Although the exact year is not known it may be the earliest note on this heavenly phenomenon. Other solar eclipses are mentioned in many other ancient Chinese documents. Statistics show that more than 1,100 eclipses were recorded.

What Causes a Solar Eclipse?

When the moon moves between the sun and the earth, it blocks the sunlight from reaching the earth, resulting in a solar eclipse.

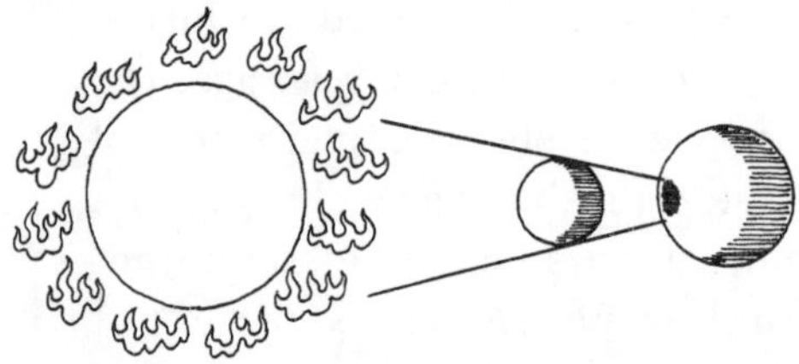

Types of Solar Eclipses

Total Eclipse
The sunlight is totally blocked by the moon.

Partial Eclipse
Part of the sun is blocked.

Annular Eclipse
The central part of the sun is covered.

The Heavenly Dog Devouring the Sun

In the ancient times, the phenomenon of the solar eclipse was called "the heavenly dog devouring the sun". To save the sun, people would bang loudly on plates to scare away the legendary dog.

Lunar Eclipses

The oracle-bone inscriptions unearthed from the Yinxu Ruins in Anyang of He'nan Province carry the earliest record of a lunar eclipse which occurred between 1300 BC and 1400 BC. Zhang Heng, an astronomer of the Eastern Han Dynasty, offered a scientific interpretation of the lunar eclipse, saying that the moon shines by reflecting sunlight. Thus, a lunar eclipse takes place when the earth moves between the sun and the moon.

What Causes a Lunar Eclipse?

When the earth is between the moon and the sun, the sunlight is blocked by the earth and cannot reach the moon. Thus, a lunar eclipse occurs.

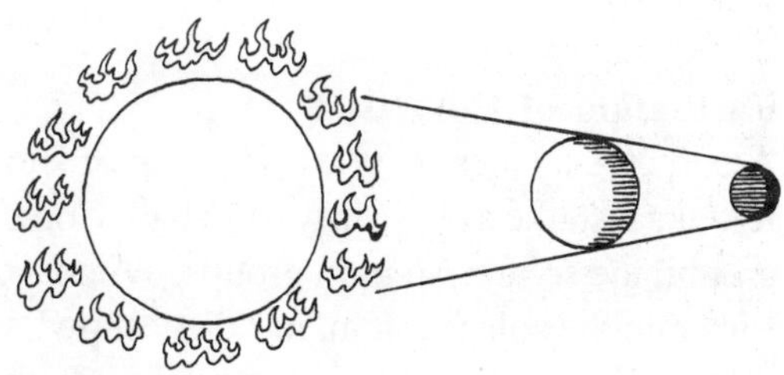

Types of Lunar Eclipses

Total Eclipse

The entire moon enters the umbra of the earth.

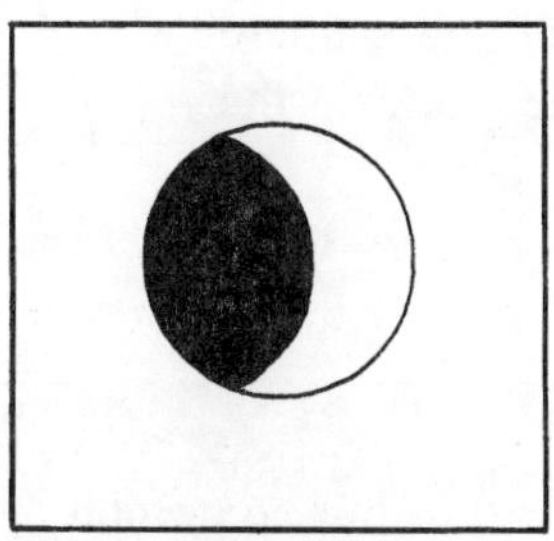

Partial Eclipse

Only part of the moon enters the umbra.

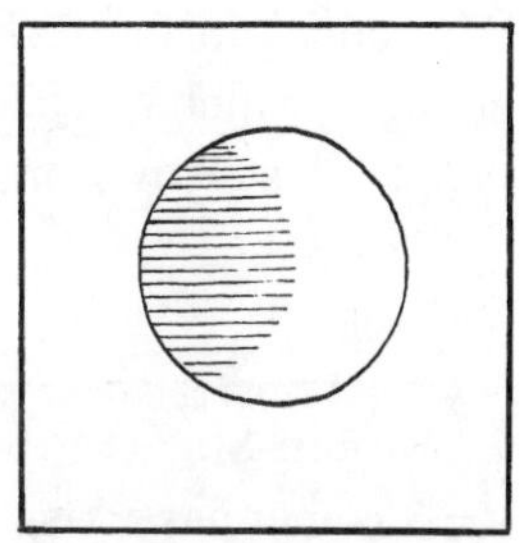

Penumbral Eclipse

The moon moves across the earth's penumbra and darkens slightly. It is quite difficult to see this with the naked eye.

Comets

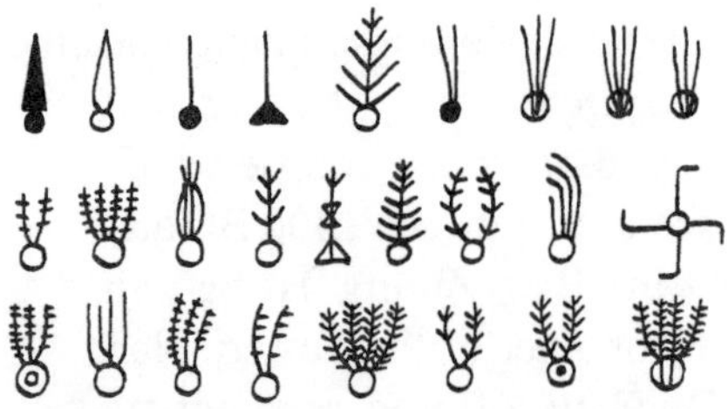

Comets have been mentioned in Chinese documents at least 500 times from the distant past to the late Qing Dynasty. These notes described not only the shapes, colours, times of appearance, positions and directions of motion of the comets, but also the rare phenomenon of comet-splitting. A silk painting with 29 comets of different appearances was discovered in a former Han tomb in Mawangdui, Changsha of Hu'nan Province. These are the oldest known pictures of comets in the world.

A note in the *Annals of the Spring and Autumn Period* describes a comet that visited the earth in 613 BC: "In July of autumn, a comet came from the Big Dipper." This is also the earliest note on comets. The oldest record in the West was in AD 66.

Formation of Comets

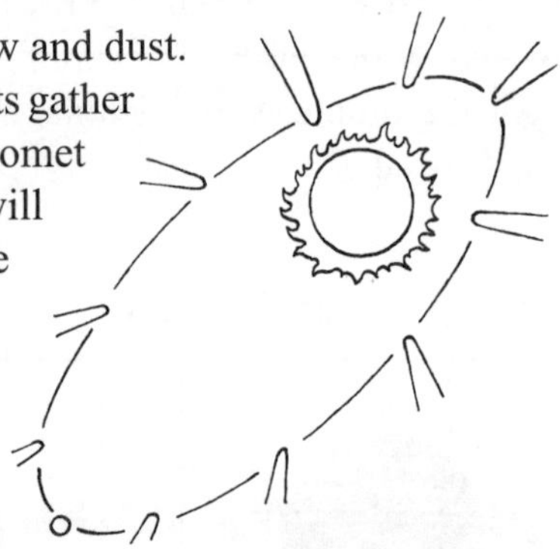

A comet is a celestial body made up of snow and dust. It looks just like a huge dirty snowball. Comets gather around the solar system in groups. When a comet flies into the solar system, its "dirty snow" will turn into gas and dust as it comes close to the sun. The gas and dust leaves a long tail behind that shines in the reflected and scattered sunlight, which will become shorter as the comet flies away from the sun.

Broom-Star (Comet)

A comet has a long tail and is thus commonly called a 'Broom-Star'. The ancients believed it was a symbol of war and calamity. Their appearances heralded wars and disasters for humankind, and ancient people were afraid of seeing them. People who bring misfortune or bad luck may also be called Broom-stars.

The Best Known Comet — Halley's Comet

This comet makes one orbit around the sun every 76 years. In 1682, the British astronomer Edmond Halley discovered that the elongated path of a comet was quite close to those recorded in 1531 and 1607. So he predicted that the comet would return to the earth 76 years later. When it returned as he had predicted in 1759, it was named Halley's Comet in his honor.

Star Charts

The oldest star chart in China was compiled by Shi Shen, an astronomer of the Warring States Period. Listing the equatorial coordinate positions of over 110 stars, the chart laid the foundation for later measurements of celestial bodies. The oldest star chart in Europe was drawn up two centuries later than Shi Shen's chart.

ASTRONOMICAL INSTRUMENTS

Armillary Sphere

Zhang Heng, the Eastern Han Dynasty astronomer, constructed the first equatorial armillary sphere, a model of the celestial sphere used to determine the position of stars.

Structure and Operation

Zhang Heng's armillary sphere was a spherical model made of steel, with the celestial sphere and the time-keeping clepsydras (water clocks) connected by gears. Powered by the water dripping from the clepsydras, the gears rotate the sphere at a daily pace synchronized with the real celestial sphere. With the aid of this armillary sphere, people could clearly observe the rising and setting of the sun, moon and stars. A mechanical calendar displaying the date was also mounted on the instrument.

Simplified Armillary Sphere

The Yuan Dynasty astronomer Guo Shoujing developed a simplified armillary sphere. As the most important astronomical instrument in ancient China, the armillary sphere included equatorial, tropic and other rings to display the coordinates of the celestial bodies. With eight or nine rings, the instrument used to operate very slowly and laboriously. Guo Shoujing dispensed with many of these rings, relocated the equatorial ring to the lower part and removed the ecliptic ring. Much easier to use, this simplified sphere had minimal errors.

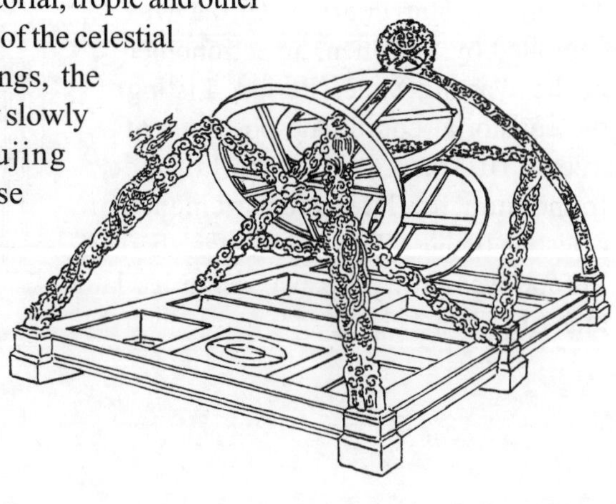

Astronomical Clock

Monk Yixing of the Tang Dynasty, Zhang Sixun and Su Song of the Song Dynasty improved on Zhang Heng's armillary sphere and created the earliest astronomical clock in the world.

The device integrates observation, demonstration and time-telling:

- The uppermost level is an observatory platform on which stands an armillary sphere;
- On the middle level is a spherical armillary sphere with 283 constellations and 1,464 stars drawn on it. The sphere rotates synchronously with the sky;
- On the lowest level are a water-powered mechanism and a well-made time-keeper which tells the time with a little puppet.

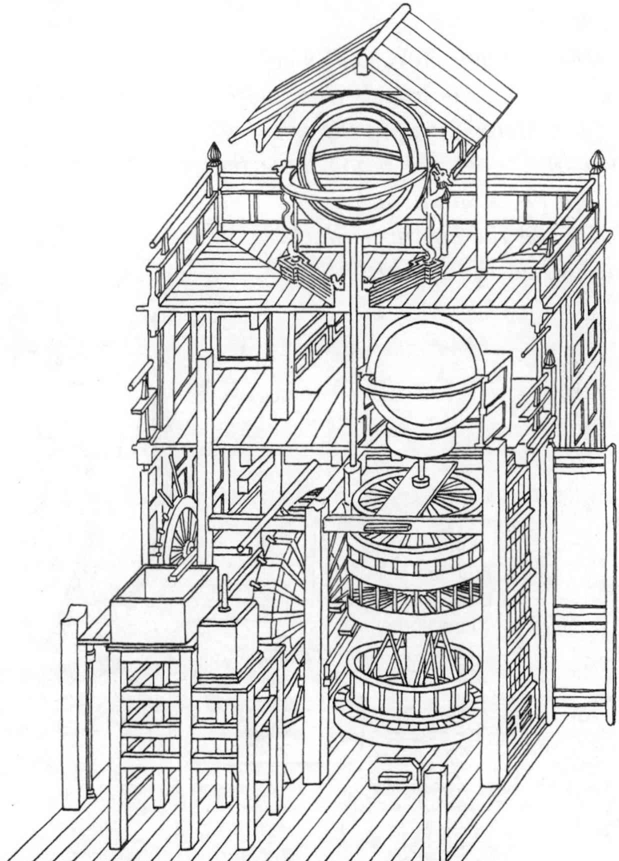

A Poem of Investigation — Qu Yuan's "Questioning Heaven"

Qu Yuan was a great Chinese poet who lived during the Warring States Period. He composed a poem entitled "Questioning Heaven", in which he raised a staggering 172 questions. Many of these questions inquire into nature and perfectly embody the ancient people's penchant for scientific exploration. Here are some of the questions he asked:

Who originated the story about the birth of the universe?
How can heaven and earth be known before they have realised their form?
What brought about the unity of yin, yang, heaven and earth so that everything on earth came into existence?
Who designed and built the nine-tiered heaven? Who was powerful enough to create it?
How do heaven and earth move and meet?
What is the distance the sun travels from the Boiling Canyon in the morning to the bank of the Meng River in the evening?
What special quality does the moon have that enables it to die and then come back to life (wane and wax)?
What is closed when the sky turns dark?
What is opened when the sky brightens?

CALENDARS

A calendar is a method of calculating time using terms like year, month and day. Calendar-making is mainly based on astronomical observations. The Chinese people started to make and use calendars a long time ago, This practice is believed to have been initiated by the Yellow Emperor.

According to many calendars adopted after the Spring and Autumn Period, a year has 365.25 days. The Yuan Dynasty scientist Guo Shoujing set out the *Time-Telling Calendar*. According to him, a year consists of 365.2425 days, a value differing from the time the earth needs to complete a revolution around the sun by only 26 seconds. This figure agrees with the current Gregorian calendar, although Guo's calendar appeared over 300 years earlier.

The various calendars

Calendars are divided into three types: solar, lunar and combined. The solar calendar year refers to the period of time the earth takes to finish a revolution around the sun. This calendar originated in Rome. The lunar calendar bases the length of a month on the time the moon requires to finish a rotation around the earth.

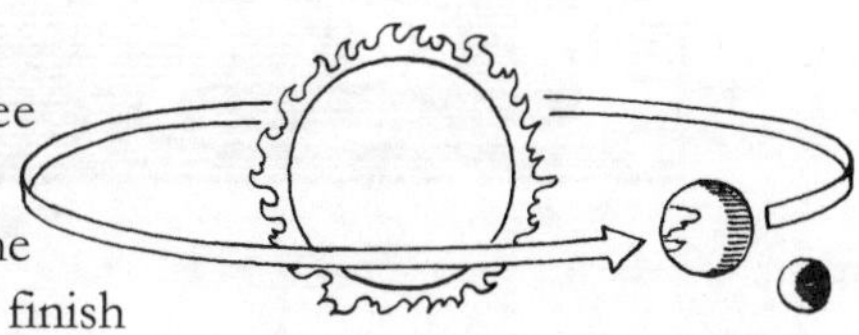

The combined calendar takes into account movements of both the sun and moon and uses the leap month to balance out the different lengths of year for the solar and lunar calendars.

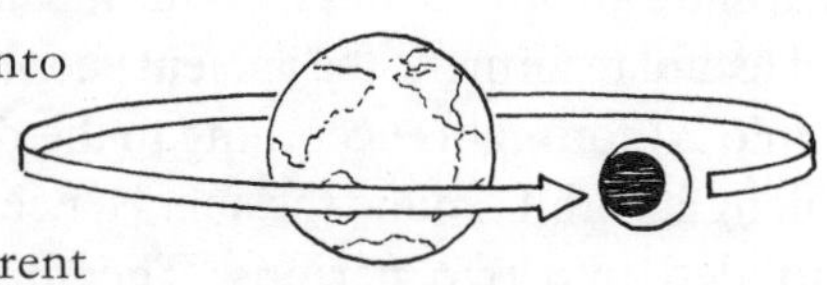

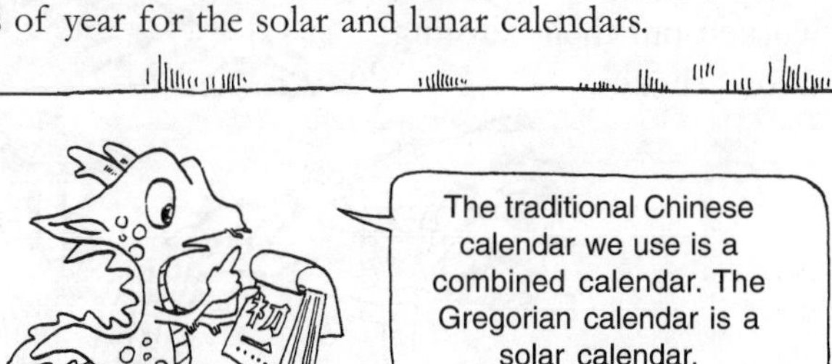

Characteristics of the Chinese Traditional Calendar

Synodic Month (*Suo Wang Yue*)

The traditional Chinese calendar is based on the synodic month, which is the time taken by the moon to make a complete circle around the Earth. The month begins with the new moon (the first day), progresses to the full moon (the 15th day), then ends with the new moon. A month has 29 (minor month) or 30 days (major month) while a year is made up of 354 or 355 days. A leap month will have to be inserted every three years.

Twenty-Four Fortnightly Periods (*Jie Qi*)

The waxing and waning of the moon does not reflect seasonal and climatic changes. As the lunar calendar was not useful to farmers, the ancient people divided a year into 24 fortnights according to the positions of the earth along its orbit around the sun. Hence, each month was divided into two periods. The farming community then worked out their farming schedules based on these periods.

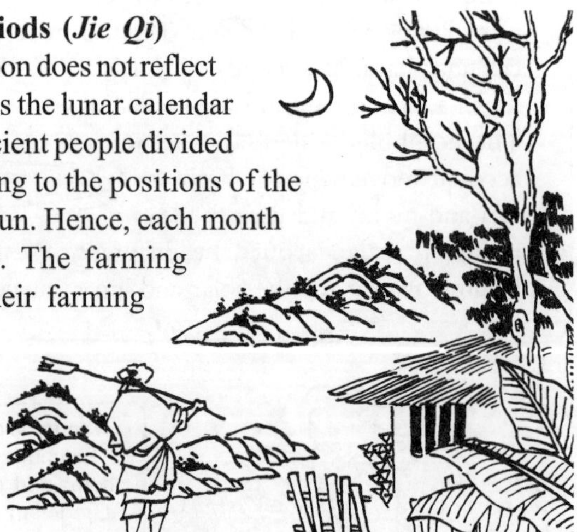

Twenty-Four Fortnightly Periods			
Spring	First month	*Li-Chun*	The spring begins and it turns warm.
		Yu-Shui	It starts to rain.
	Second month	*Jing-Zhe*	Thunderstorms come; hibernating animals become active again.
		Chun-Fen	In the middle of spring; day and night are equal in length.
	Third month	*Qing-Ming*	Warm; weather is clear and bright.
		Gu-Yu	Rainfalls increase; good for planting crops.
Summer	Fourth month	*Li-Xia*	Summer begins; temperatures rise.
		Xiao-Man	Seeds of wheat and other crops maturing in summer become plump.
	Fifth month	*Mang-Zhong*	Wheat and other awny crops mature.
		Xia-Zhi	Summer arrives. Longest days and shortest nights of the year.
	Sixth month	*Xiao-Shu*	It is becoming hot.
		Da-Shu	The hottest period of the year.
Autumn	Seventh month	*Li-Qiu*	Autumn begins and temperatures fall.
		Chu-Shu	The summer heat will pass soon.
	Eighth month	*Bai-Lu*	The night is cool and dewdrops appear.
		Qiu-Fen	In the middle of autumn, day and night are equal in length.
	Ninth month	*Han-Lu*	Temperatures drop significantly and dewdrops are cooler at night.
		Shuang-Jiang	Frost starts.
Winter	Tenth month	*Li-Dong*	Winter begins and the weather turns cold.
		Xiao-Xue	Snowfall starts.
	Eleventh month	*Da-Xue*	Heavy snow appears.
		Dong-Zhi	A period of severe cold starts. Shortest days and longest nights of the year.
	Twelfth month	*Xiao-Han*	The weather turns much colder.
		Da-Han	The coldest period in the year.

Heavenly Stems and Earthly Branches

The traditional Chinese calendar uses heavenly stems and earthly branches to count years:

The 10 Heavenly stems: *Jia, Yi, Bing, Ding, Wu, Ji, Geng, Xin, Ren, Gui.*

The 12 Earthly branches: *Zi, Chou, Yin, Mou, Chen, Si, Wu, Wei, Shen, You, Xu, Hai.*

These stems and branches combine to form 60 groups representing 60 years, a period called *Jia-Zi*. The cycles of *Jia-Zi* are used by the traditional Chinese calendar to count years, such as *Jia-Zi*, *Yi-Chou*, *Bing-Yin*, etc.

TIMEKEEPERS

Measuring the Sun's Shadow

In the ancient times when there was no clock, people started to work at sunrise and went to bed at sunset. The rising and setting of the sun were used to indicate the time for work and rest. The length of a shadow cast by an upright pole could be used to tell the time.

Sunlight and Shadow

Objects have shadows when they stop sunlight from reaching the ground.

The length of the shadow varies with the position of the sun at different times of the day.

At noon, the sun is directly overhead and only a small portion of its light is blocked by the object. Hence, its shadow is short at this time.

In the morning or evening, the sun is at a lower angle and a large proportion of its light is blocked by the object. This results in a longer shadow.

Sundial

A sundial consisted of a gnomon and a stone base-plate. On the plate were time markings and the shadow of the gnomon served as the hour hand. Thus it was possible to determine the time to some extent.

Clepsydra (or Drip Vessel)

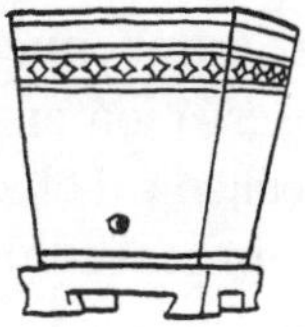

Perhaps people long ago might have broken pots by accident, causing leaks. They would have discovered that the water level in the pot went down at a regular rate as time passed. For instance, half of the water would take half an hour to leak out, while the remaining part would take another 30 minutes. So it was used as a time-keeper and the clepsydra was invented.

The earliest clepsydras were just pots with a hole drilled at the bottom. They were filled with water and gauze was placed on the water surface to indicate the water level as it went down. However, this pot needed to be refilled again and again, and its accuracy in keeping time was thus compromised.

Later, Zhang Heng of the Eastern Han Dynasty put two pots together. The water dripped out of the first pot into the second, which had a marked gauge that went up with the water level and showed the time.

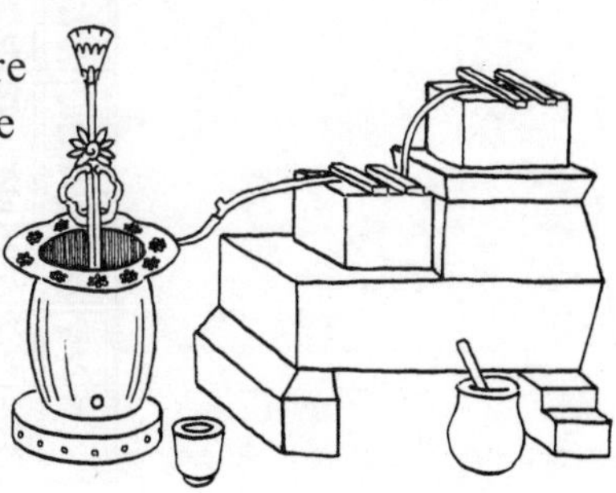

Over time, the number of pots used were increased and the pots were set up in multiple tiers. During the Song Dynasty, a lotus clepsydra appeared. It was able to keep time extremely accurately, with just an error of 20 seconds per day. Its accuracy remained unmatched until the invention of mechanical clocks in the 18th century.

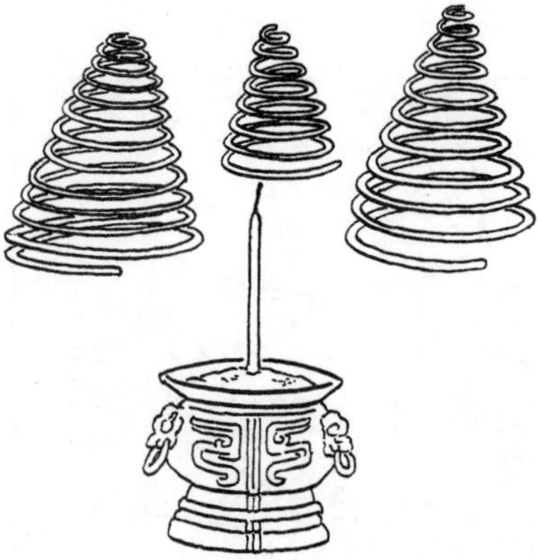

Incense Clock

Sticks of incense with markings were also used to keep time. They were made in various shapes, with some of them even twisted into the Chinese characters for Happiness, Longevity and Joy, etc. Some of these incense sticks were very long and might take over ten days to be burnt out.

Mechanical Clocks

The astronomical clock previously mentioned, which was used for both heavenly observation and telling the time, is actually a mechanical clock. The earliest mechanical clock in the world was made by Monk Yixing of the Tang Dynasty. He improved the armillary sphere designed by Zhang Heng by mounting two puppets. One of the puppets would beat the drum every 15 minutes while the other would strike the bell every two hours. It was like an alarm clock and was driven by water. Su Song of the Song Dynasty made a more elaborate mechanical clock. Twelve metres high, the clock was divided into seven levels with over 160 wooden puppets announcing the hour.

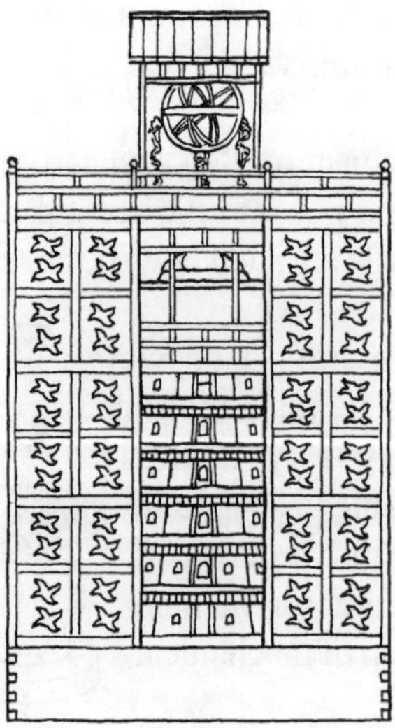

A Riddle: How Far Is the Sun From Us?

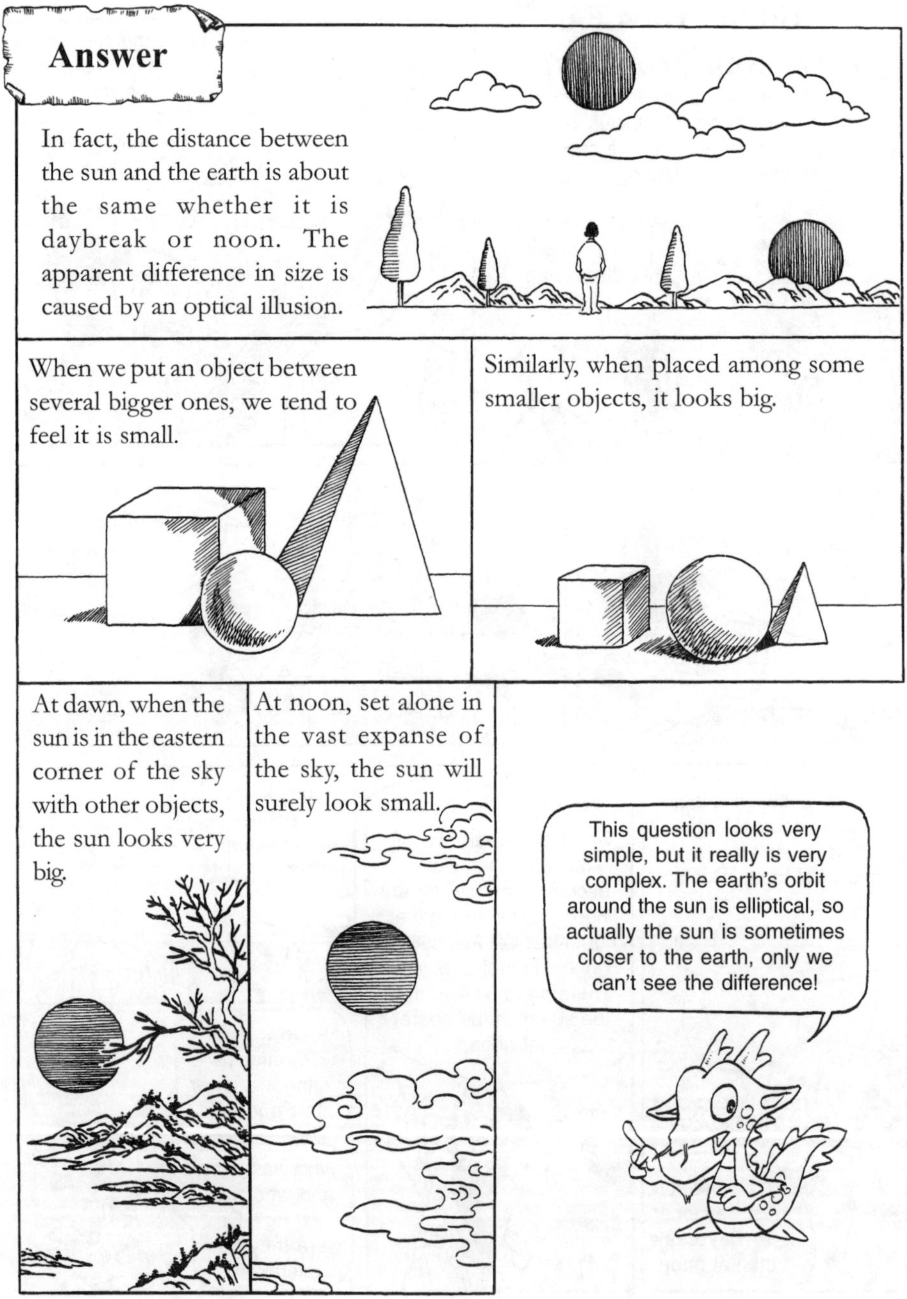
Answer
In fact, the distance between the sun and the earth is about the same whether it is daybreak or noon. The apparent difference in size is caused by an optical illusion.
When we put an object between several bigger ones, we tend to feel it is small.
Similarly, when placed among some smaller objects, it looks big.
At dawn, when the sun is in the eastern corner of the sky with other objects, the sun looks very big.
At noon, set alone in the vast expanse of the sky, the sun will surely look small.
This question looks very simple, but it really is very complex. The earth's orbit around the sun is elliptical, so actually the sun is sometimes closer to the earth, only we can't see the difference!

GEOGRAPHY

China has also scored outstanding achievements in the field of geography, boasting the first seismograph and the earliest scaled maps in the world. Prominent geographers Xu Xiake and Li Daoyuan have both made significant contributions to China's study of geography with their classic works.

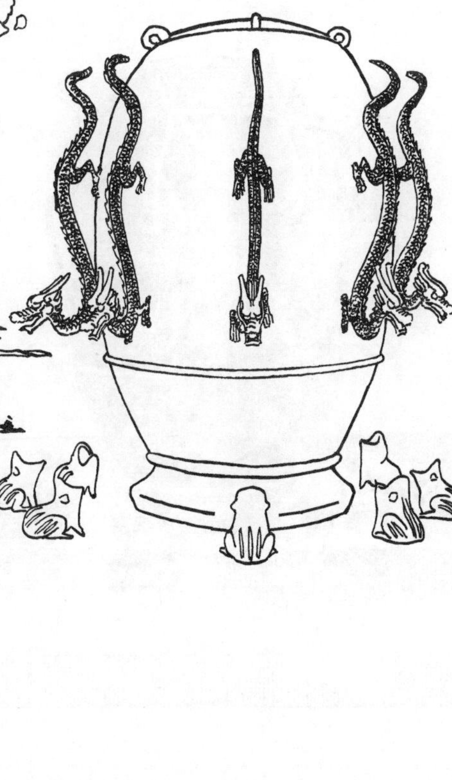

SEISMOGRAPH

The Ancient Seismograph

The Eastern Han Dynasty scientist Zhang Heng invented the first seismograph in the world in AD 132.

Structure and Operation

The seismograph was shaped like a big wine goblet with a round lid, and inscribed with mountains, tortoises, birds, beasts and characters on its surface. It is encircled by eight dragons with copper balls in their mouths looking down at eight copper squatting toads, each facing a different direction. A copper pole in the middle is attached to eight connecting rods. When there is a seismic tremor, the pole would bend. This will cause the dragon facing that way to drop its ball into the mouth of the toad, showing the direction of the earthquake.

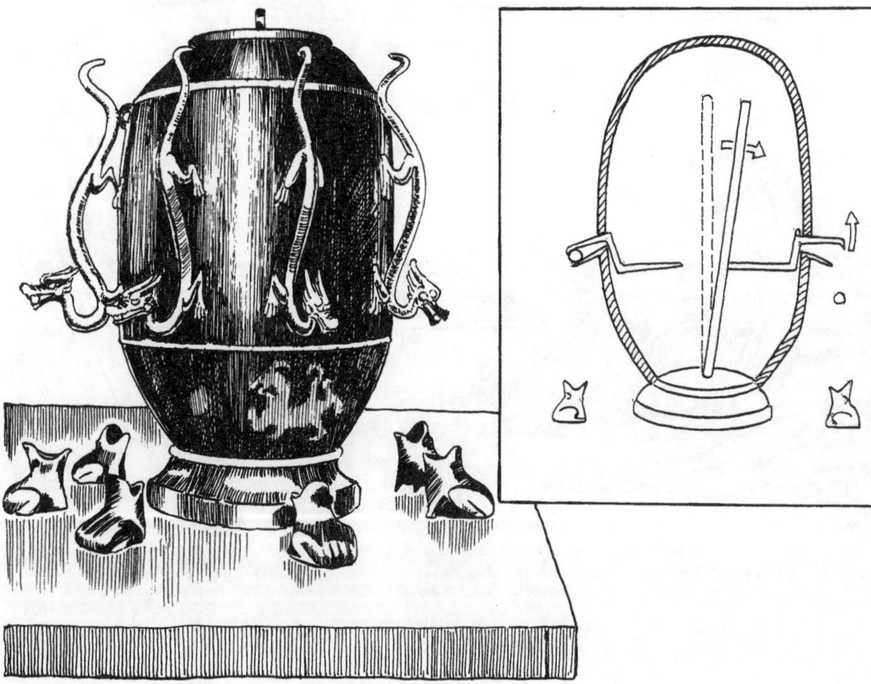

MAPS

The Earliest Scaled Maps

Forefathers of the Chinese people started to draw maps on objects more than 4,000 years ago. The Nine Bronze Cauldrons (*ding*), which were reportedly cast during the Xia Dynasty, differentiated various places with pictures of landscapes, plants and beasts unique to them. These pictures could act as a guide for people who made a journey to distant places.

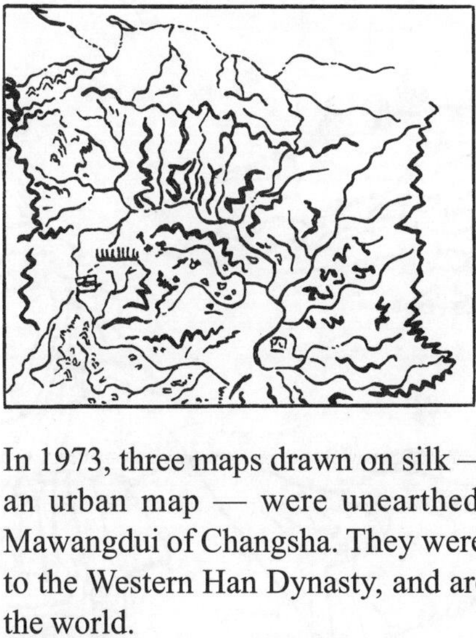

Ancient Chinese literature contains a considerable amount of information on maps. There were many types of maps for various purposes such as administrative divisions, agricultural land use, topography, distribution of mineral resources, roads and transportation, tomb locations, and military affairs.

In 1973, three maps drawn on silk — a topographical map, a garrison map and an urban map — were unearthed from the No.3 Han Dynasty Tomb in Mawangdui of Changsha. They were drawn about 2,000 years ago, dating back to the Western Han Dynasty, and are among the earliest scaled maps found in the world.

Pei Xiu

Pei Xiu (AD 224–271) is the pioneer of scientific cartography or map-making. The Six Principles he proposed stressed the importance of accurate scale, position and distance. All maps were charted in accordance with his principles in China for the next millennium. (Six principles: scale, azimuth, distance, relative height, slope, curved and straight lines.)

GEOLOGY

The Song Dynasty scientist Shen Kuo discovered the fossils of conches, oyster shells and sea-urchins during his tour of Taihang Mountain. So he concluded that although the place was thousands of miles away from the sea, it was once covered by the sea, a long time ago. The sea bottom had been elevated gradually over the years to become the mountain he saw.

In Europe, it was only in 1785 that James Hutton, the father of modern geology, came to the same conclusion in his Theory of the Earth.

Outstanding Chinese Geographers

Xu Xiake (1586–1641)

A prominent traveller and geographer of the Ming Dynasty, *The Travelogue of Xu Xiake* is the record of the observations he made on his travels between 1613 and 1639. A detailed record of both topographical and geological features and the plants he saw, the book pointed out a new direction for Chinese geographical studies by emphasising the observation and description of nature. Many geographical concepts used by later European geographers can be found in this travelogue. Xu Xiake is the first scientist in the world to conduct an extensive and systematic investigation into karst geomorphology. His work in this field anteceded the Europeans by over a century.

Li Daoyuan (466 or 472–527)

A distinguished geographer who lived in the middle and later part of the Northern Wei Dynasty, he authored the *Commentary on the Waterways Classic*. Not only does the book describe the origins, paths, changes and water irrigation projects of the Yellow River, the Yangtze River and many other rivers, it also offers detailed records of the landscapes, cities and towns, products and customs of those river basins. Thus, it is a treasured sourcebook for the study of ancient Chinese geography.

The Story of Xu Xiake

Meteorology

Weather Forecasting

The ancient Chinese had a keen interest in weather forecasting. Many of the Chinese oracle inscriptions unearthed were concerned with weather. A piece of oracle bone unearthed in the Yin ruins in Anyang of He'nan Province, from the reign of Emperor Wending of the Shang Dynasty, contains a complete record of the weather conditions from the 20th to 29th of the third month in 1217 BC. Special officials were assigned to observe and forecast wind and cloud movements in ancient times. In addition, there were many books on weather forecasting.

Proverbs on the Weather

Hardworking farmers have gathered a wealth of knowledge on the patterns of weather fluctuation in the course of agricultural production, and this knowledge is often expressed in proverbs.

Ant's nest sealed, storm approaching.		Rosy clouds at dawn — stay at home. Rosy clouds at dusk — go a thousand miles.	
Spider spins web, rains are ending.		Thick frost heralds fine weather.	
Northeast wind blows, heavy rain comes.		A setting sun as red as rouge, rain or wind will come.	

Rain Gauge

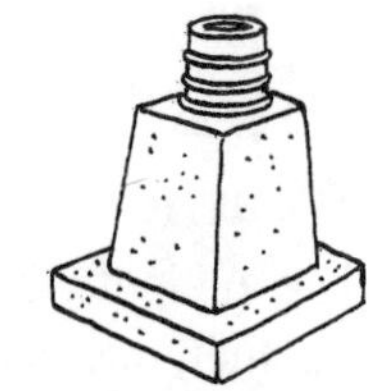

The first use of the rain gauge was reported in China during the Southern Song Dynasty. In 1424, the government made standardised rain gauges and distributed them to districts and counties across the country, requesting regular reports on rainfall. Rain gauges quickly spread to neighbouring countries, but were not used in Europe until the 17th century.

Make Your Own Rain Gauge

1. Cut an empty plastic bottle into half, invert the top half and place it in the lower half.
2. Place the rain gauge outdoors and secure it with stones.
3. Measure the rainfall and then pour out the rainwater. Do this daily.

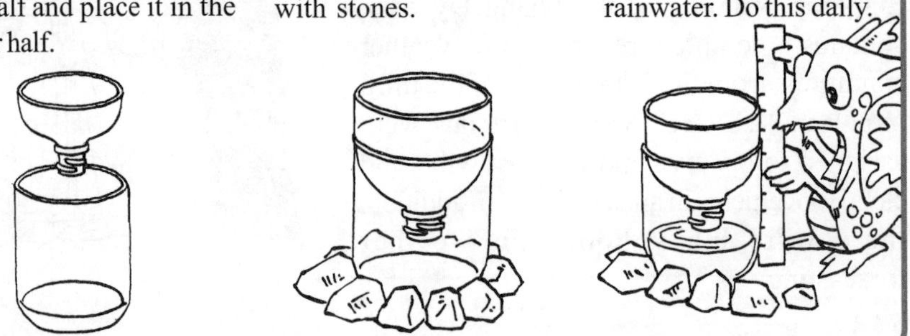

Watching the Clouds

Master Lu's Spring and Autumn Annals (*Lu Shi Chun Qiu*), written in about 239 BC, divided clouds into four types: mountain cloud, water cloud, drought cloud and rain cloud. The earliest nephogram (cloud atlas) found so far is the *Astronomical and Meteorological Divination* drawn in the Western Han Dynasty. Dynasties thereafter all kept a large number of nephograms.

Cloud and Weather

When broom-like clouds appear in the sky, the ground will be wet with rain.

If pagoda-like clouds are seen in the morning, it will rain cats and dogs in the afternoon.

If white clouds pile up like a fortress, a thunderstorm is imminent.

Clouds like broken stones will bring unbearably hot temperatures.

Clouds like steamed buns foretell fine weather.

Dark clouds come before rain.

It is windy when the cloud head is dark; it rains when the cloud head is white.

If the dark clouds meet the sun at a high position, it will rain the next morning.

If the dark clouds meet at a low position, it will rain in the evening.

Wind Vane

Zhang Heng of the Eastern Han Dynasty made a copper bird with three feet and put it on a bamboo pole. It could turn when the wind blew to show the direction of the wind. This was 1,000 years before Europe's weather vane.

The *Yi-Si Divination* by Li Chunfeng, a Tang Dynasty official in charge of astronomy and the calendar, is the earliest meteorological monograph. In it, he classified the wind into eight scales and 24 directions. Four centuries later, the book was taken to Europe and British scholars further divided the wind velocity (speed) into 13 scales ranging from zero to 12.

Make Your Own Wind Vane

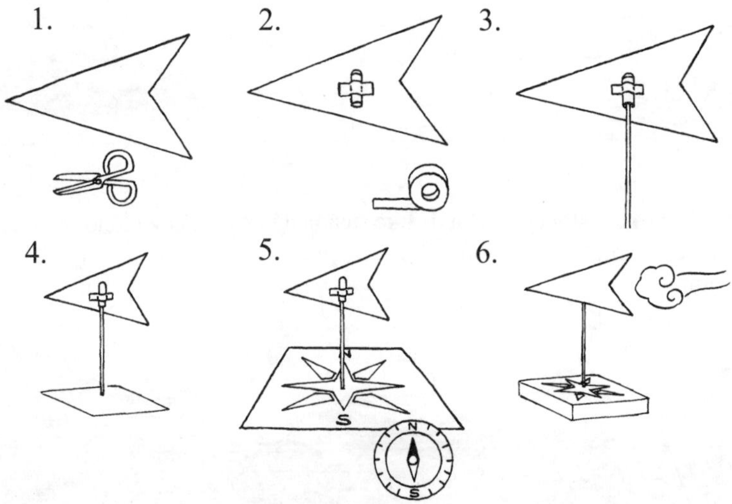

1. Cut an arrow-like marker from cardboard.
2. Attach a pen cap to the tail of the marker.
3. Get a long needle and insert it into the cap.
4. Stick a piece of paper onto a base and poke the needle with the marker in.
5. Use a compass to find the directions and indicate them on the base.
6. When the wind blows, the marker will point to the direction it comes from.*

*The larger vane tail causes stronger resistance to the wind. Only when the part of the vane tail facing the wind turns to a position parallel to the wind direction can its resistance be minimised. So, the vane marker will point at the direction the wind is coming from.

The Shape of Snowflakes

The Chinese people had already discovered this in the second century BC. It was not until 1611 that the Europeans finally made the same discovery. Han Ying of the Han Dynasty wrote in *Han's Comments on the Book of Poetry*: "Most grass and trees have pentagonal flowers while snowflakes are mainly hexagonal. So the latter is *yin* in property like water."

For the Chinese people, the number six is associated with water: Six is the multiple of water according to the Chinese classic *The Book of Changes* (*I Ching*). Snowflakes are crystallised flowers of condensed water; thus, they are hexagonal.

Ancient Phenological Records

The ancient Chinese recognised the relationship between periodic changes in the climate, the growth of plants and the activities of animals. The earliest known phenological record is a book called *Xia Xiao Zheng* (*Lesser Annuary of the Xia Dynasty*), written 3,000 years ago. An entry on the climate during the first month of the lunar calendar reads: "The hibernating insects are waking up, the wild geese flying northward, the wild chickens fluttering their wings and crying, water temperature rising, fish swimming upstream… plums, apricots and peaches blooming…"

Zhuge Liang Borrows the East Wind

In the literary classic, *Romance of the Three Kingdoms,* there is a wonderful story of how Zhuge Liang used the east wind to defeat Cao Cao's army.

Mathematics

Like the other fields of science, mathematics also developed very early in China. China is the first civilisation in the world to adopt the decimal system and the binary system. The multiplication table and the order of the four operations were also established by China. The Chinese also invented the abacus, a simple yet effective tool for calculation that is comparable to computers.

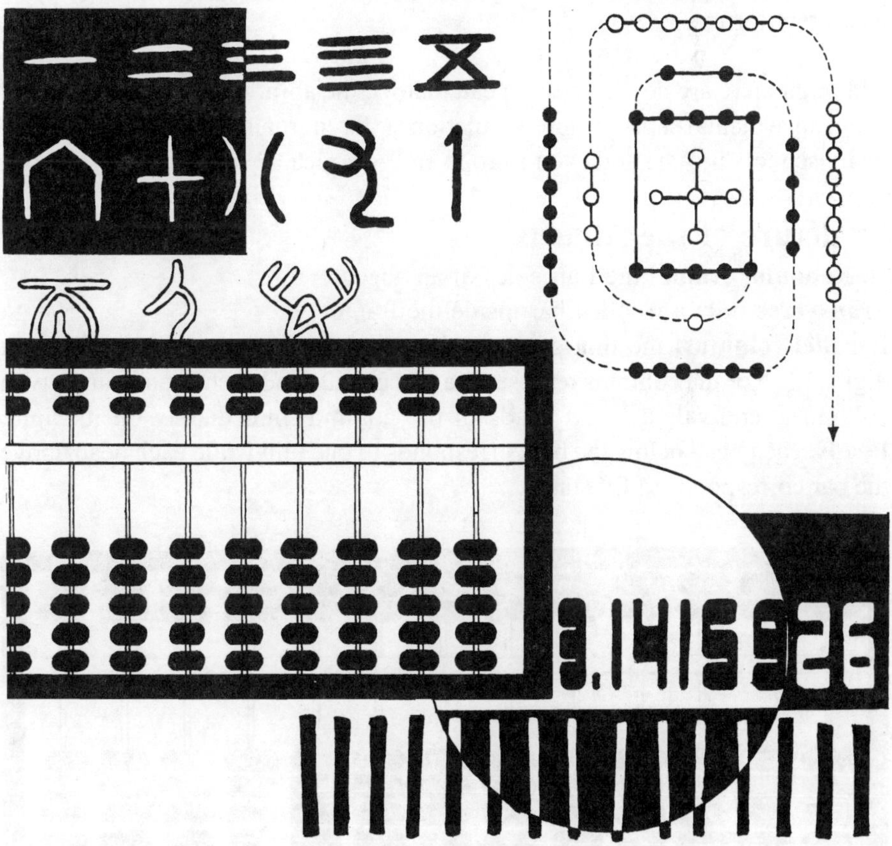

Abacus

The abacus is a traditional tool for calculation. Reckoning numbers using the abacus is known as bead arithmetic.

The first mention of bead arithmetic occurs in the *Memoir on the Methods of Numbering* (*Shushu Jiyi*) written over 1,700 years ago during the Eastern Han Dynasty. The earliest known drawing of an abacus appears in the Northern Song painter Zhang Zeduan's masterpiece, *A Trip Up the River During Qingming* (*Qing Ming Shang He Tu*). By that time, the abacus was already in its current form.

Although there are now electronic calculators, the ability of the ancient abacus to enhance mental capacity and calculation skills continues to ensure its place in many schools in Asia and even Europe and America.

Structure of the Abacus

Rectangular frame: the four sides of an abacus.
Transverse bar: a wooden bar inside the frame.
Parallel columns: the thin columns threading the bar are used to represent digits. Each of the columns represents a place value and each bead in any given column is equivalent to 10 beads in the column immediately on its right.
Beads: each bead below the bar corresponds to one unit while each bead above the bar corresponds to five units.

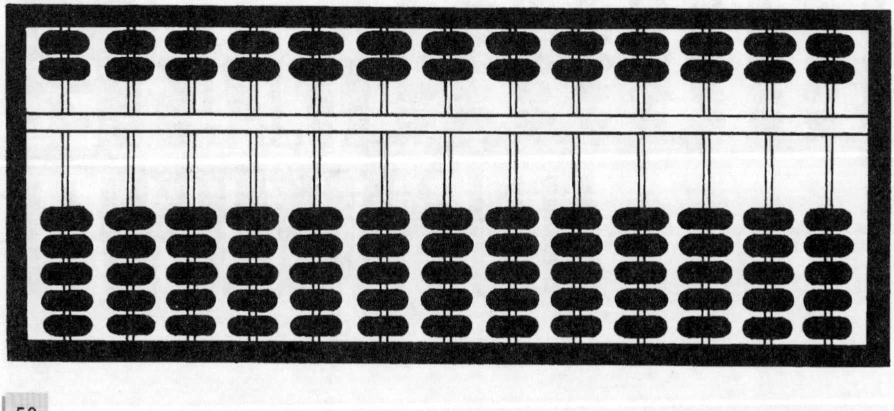

Types of Abacus

There are various types of abacus, with different numbers of columns and arrangements of beads.

The traditional Chinese abacus features wide spacing and drum-like beads, two above and five below.

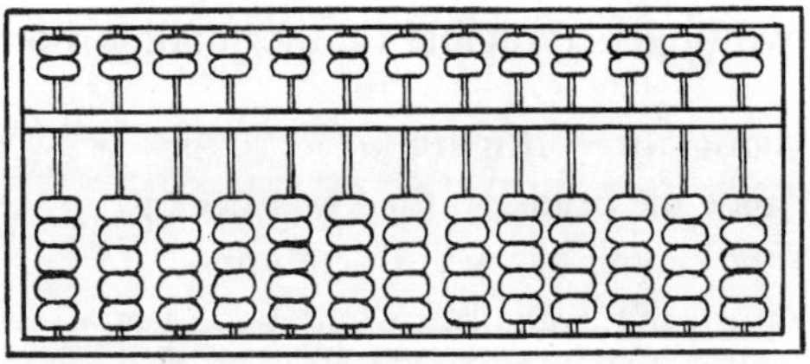

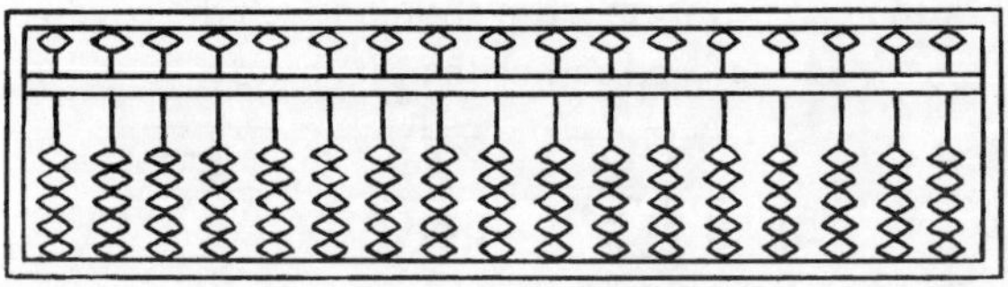

The Chinese abacus was brought to Southeast Asia during the Ming Dynasty. In the middle of the 17th century, the Japanese modified it to come up with a Japanese version with narrow spacing and rhombic (diamond-shaped) beads. There are three types of Japanese abacus: two above and five below; one above and five below; and one above and four below. The latter two are more popular.

There is also a medium-sized abacus with moderate spacing and dish-like beads, one above and four below.

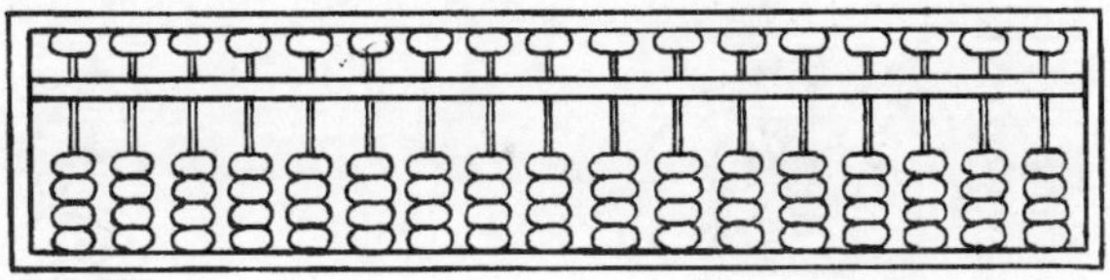

Other uses of the abacus

Besides calculation, the abacus can also be used by fortune-tellers to foretell distant or imminent misfortunes. A huge abacus is usually hung in a temple of the town god to represent judging persons on their merits and faults. The Ming Dynasty book on the abacus, *Systematic Treatise on Arithmetic* (*Suan Fa Tong Zong*) includes a mnemonic for predicting the gender of a child.

How to Use the Abacus

The abacus is easy to learn — all you need is a basic knowlege of how to move the beads efficiently and how to calculate sums. You can move the beads with three fingers (suitable for a larger abacus with seven beads in each column) or two fingers (suitable for the smaller rhombic abacus).

Using three fingers:

- slide the beads on the lower deck to the central bar with the thumb.
- push the beads below the bar away from the bar with the index finger.
- move the beads above the bar with the middle finger.

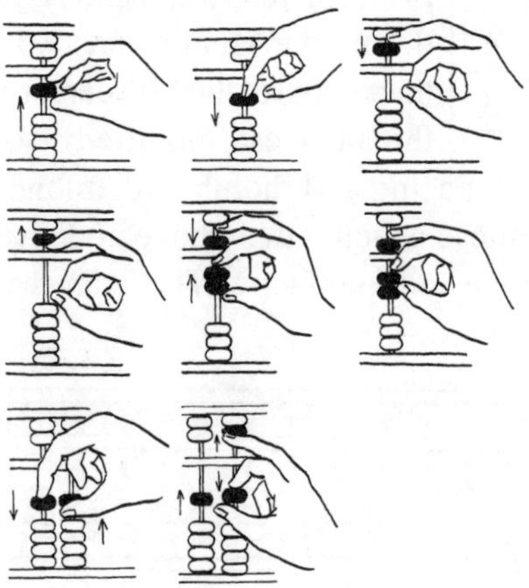

Using two fingers:

- use the thumb to move the beads below the central bar.
- use the index finger to move the beads above the central bar.
- your fingertip should touch the edge of a bead rather than its centre.

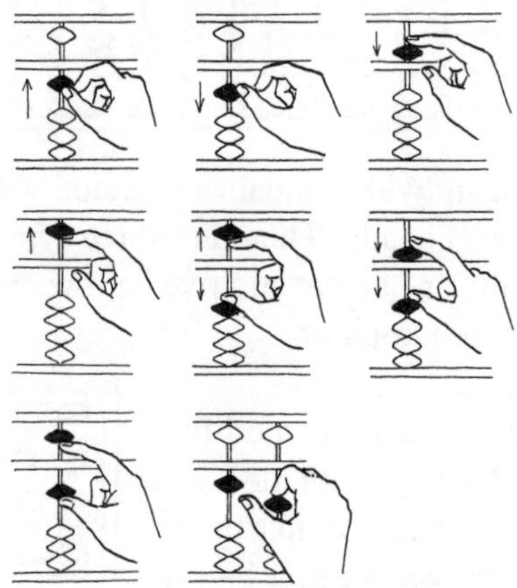

Example: 2316 + 1163

1. Set the beads to show the number 2316.
   (a bead above the bar represents five units)
2. — Add 1 to the thousands digit by moving a bead below the bar to the bar;
   — Add 1 to the hundreds digit by moving a bead below the bar to the bar;
   — Add 6 to the tens digit by sliding a bead above the bar (5) to the bar and a bead below the bar (1) to the bar;
   — Add 3 to the ones digit by moving three beads below the bar to the bar.
3. The answer is 3479.

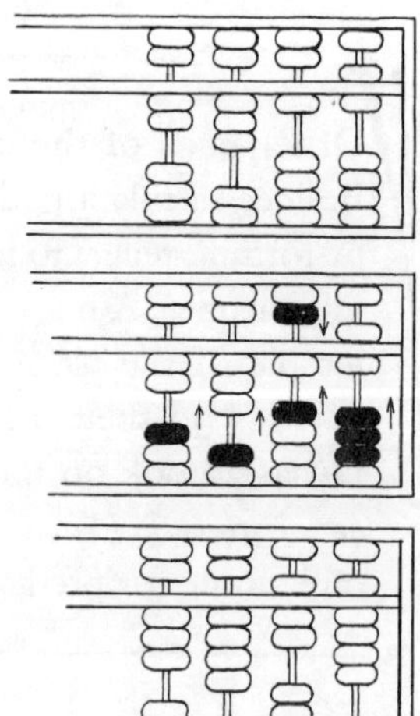

Before the Abacus — Counting Rods

Before the invention of the abacus, the ancient people used counting rods for calculation, as early as the 5th century BC. Counting rods were small sticks of equal length, made of bamboo, wood, iron, jadestone, ivory, etc.The rods were arranged according to certain rules to represent numerals, using alternating vertical and horizontal arrangements:

𝍠 𝍡 𝍢 𝍣 𝍤 𝍥 𝍦 𝍧 𝍨

𝍩 𝍪 𝍫 𝍬 𝍭 𝍮 𝍯 𝍰 𝍱

To represent multi-digit numbers, the ones were represented by vertical rods, the tens by horizontal rods, the hundreds becoming vertical again and the thousands horizontal, etc.

For instance, the year 1994 is 𝍩𝍨𝍱𝍣

The counting rod system left much to be desired. The more digits there were involved, the bigger the space required. In addition, when the calculations were performed at an increased speed, mistakes in the form of misarranged rods became more common. So, the counting rods gave way to the abacus, which used beads fixed in a frame.

Other Mathematical Achievements

Decimal System

China was the first country in the world to adopt the decimal system. Nine basic symbols corresponding to the numerals from 1 to 9 can be found in the oracle and cauldron inscriptions made during the Shang Dynasty and the Western Zhou Dynasty. There are also characters representing the tens, hundreds, thousands and ten thousands.

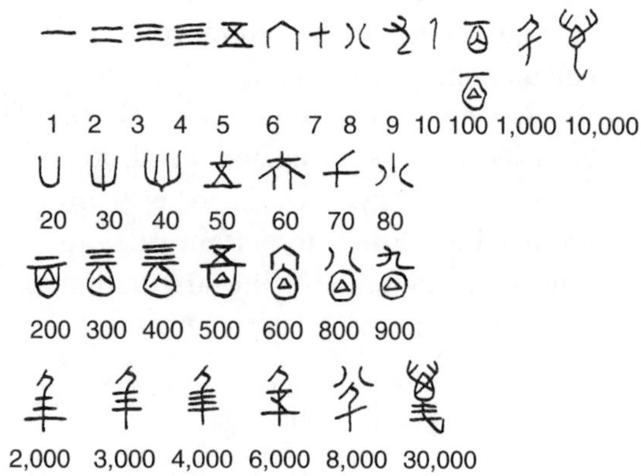

The Concept of Zero

The concept of zero is an important Chinese invention. In 400 BC, the Chinese started to use a blank space for 0. The blank is used to represent 0 both in counting rods and the abacus. The prototype of 0 was a square (□). For ease of writing, it gradually became a circle. Later, the method of using a blank for 0 spread to Cambodia, from which it went to India and other places across the world.

Negative Numbers

As early as the 2nd century BC, China had already recognized and begun to use negative numbers. They believed that the incoming and outflow of the same amount of money meant different things. So they used red to indicate positive numbers and black to indicate the negative numbers. Rules were also developed to govern the addition and subtraction of negative numbers, eight to nine centuries before the West.

Multiplication Table

The earliest mathematical table in Chinese history is the multiplication table. The table was already in general use during the Spring and Autumn Period. It also appeared on some bamboo slips from the Western Han Dynasty.

Nine Chapters on the Mathematical Art (Jiu Zhang Suan Shu)

This is the oldest mathematical book available in China. Comparable in significance to *The Elements* by Euclid of ancient Greece, the book exerted great influence on the development of mathematics in the ancient times. Written in the late first century AD (the Eastern Han Dynasty), the *Nine Chapters* is quite extensive in scope, covering most of the arithmetic, algebra and geometry in modern elementary mathematics. Among other things, the concept of fractions and their operations, calculation of ratios, introduction of negative numbers, and the rules on addition and subtraction of both positive and negative numbers were all laid out in the book, 800 years earlier than in India and at least 1,000 years before Europe. The book was brought to Korea and Japan during the Sui and Tang Dynasties and designated as a standard textbook. So far, it has been translated into Japanese, English, Russian, German, French and many other languages.

Pi

Pi is a constant needed for operations related to the circle or sphere. Prior to the Han Dynasty, the value of *pi* was defined simply as three. Liu Hui, a mathematician of the Wei and Jin Periods, developed the art of cutting the circle and concluded through calculation that the value of *pi* was 3.1416. Zu Chongzhi, a mathematician living in the Southern and Northern Dynasties, carried on the study of *pi* and defined it as a value between 3.1415926 and 3.1415927. That was the first time the world had ever achieved a determination of *pi* accurate to seven decimal places. This record was finally broken only in the 16th century by Arabian mathematicians.

The Art of Cutting a Circle

By inscribing a polygon in a circle, it is possible to calculate the approximate area of the circle. The more finely the circle is divided, the more accurate the approximate value of *pi* becomes.

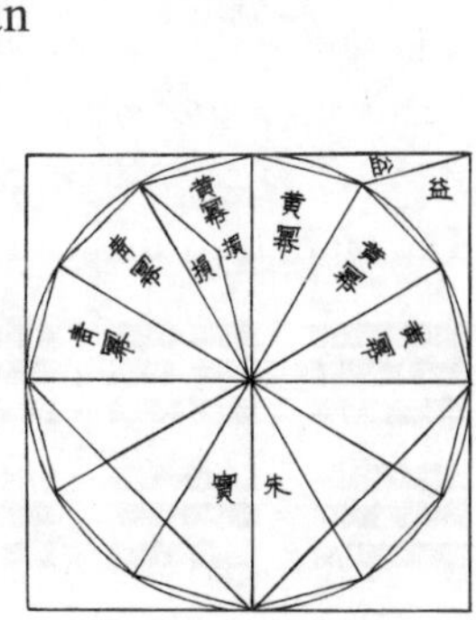

The Mathematical Theory of the Eight Trigrams

It is said that Fuxi created the mysterious Eight Trigrams in the distant past. Everything in the world can be represented by these trigrams.

The Eight Trigrams are eight symbols, which are *qian, kun, zhen, xun, kan, li, gen and dui*. Each symbol is represented by a combination of —— (full line) and – – (broken line) in its three-part structure: the top line, the middle line and the bottom line. —— is a positive or *yang* line while – – is a negative or *yin* line.

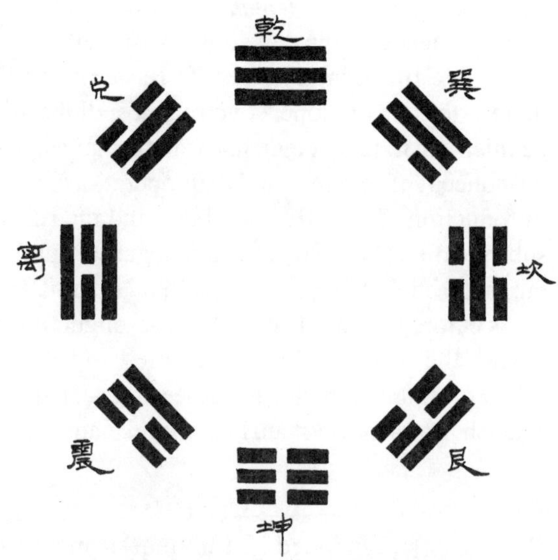

Each trigram represents something: *qian* for heaven, *kun* for earth, *zhen* for thunder, *xun* for wind, *kan* for water, *li* for fire, *gen* for mountain and *dui* for lake.

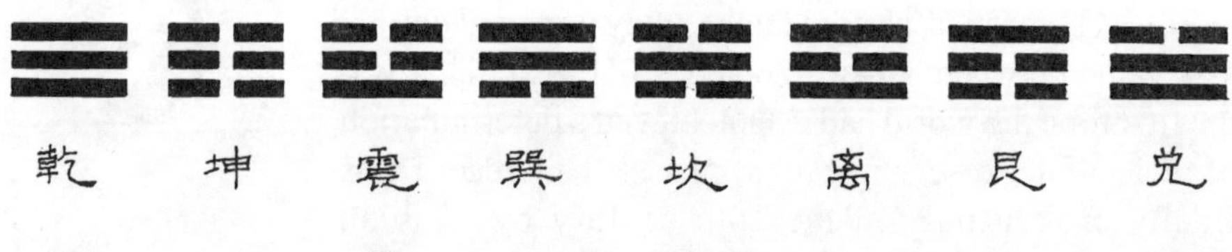

The eight trigrams can be combined to form 64 hexagrams.

In 1701, the German scientist Gottfried Wilhelm von Leibniz (1646-1716) discovered that the Eight Trigrams had elements in common with the binary system.

The Eight Trigrams and the Binary System:
The yang line ⚊ is 1 and the yin line ⚋ is 0.
The top line corresponds to the number in the first digit.
The middle line corresponds to the number in the second digit.
The bottom line corresponds to the number in the third digit.

Trigram	**Symbol**	**Binary System**	**Decimal System**
Qian	☰	111	7
Dui	☱	011	6
Li	☲	101	5
Zhen	☳	001	4
Xun	☴	110	3
Kan	☵	010	2
Gen	☶	100	1
Kun	☷	000	0

A Fool Studying Numbers

Agriculture and Industry

The ancient Chinese developed many advanced iron and bronze casting techniques. It is also the first country in the world to engage in coal mining on a large scale. China also led the world in the field of agriculture with its many efficient inventions and advanced production techniques.

Metallurgy

Iron and Steel

The forefathers of the Chinese people had already started to use iron tools during the Shang Dynasty more than 3,000 years ago.

The discovery of ironworking techniques was of revolutionary significance to China. Iron ploughs and hoes appeared in farming, and iron knives and swords were used in battles.

In the second century BC, a method was developed to convert cast iron into steel. Cast iron is rather brittle as it contains carbon. Steel is obtained by removing most of the carbon from the cast iron, making it stronger.

Frying Steel

The cast iron is heated into a molten or semi-molten state while air is pumped over it or iron ore powder is scattered onto it as it is stirred endlessly as if frying a dish. The iron ore powder and the oxygen in the air will decarburise the cast iron and turn it into steel.

High-quality steel cannot be obtained without repeated heating and forging. Hence, the Chinese proverb *bai lian cheng gang*, referring to the strength that comes after continual training.

The Chinese also invented the technique of quenching to harden steel. Specifically, the technique means plunging fiery-red iron into water to cool it. This allows the internal structure of the hot steel to be retained.

Bellows

Another important invention by the ancient Chinese people, this tool is used to remove carbon by adding oxygen to the molten iron.

The Story of Gan Jiang and Mo Ye
Gan Jiang and Mo Ye are the names of two legendary swords forged by the renowned swordsmith Gan Jiang during the Spring and Autumn Period.
One day, the King of the State of Wu ordered him to make some swords. As he and his wife Mo Ye were trying to melt the iron in a blast furnace, the temperature dropped suddenly. They could not melt the metal.
I was told that someone has to be sacrificed in order to end a mysterious phenomenon like this. Will one of us have to do that today?
My master also encounted this phenomena once. The iron would not melt until he and his wife jumped into the furnace.
At this, Mo Ye cut her hair off and pulled out her nails, and threw herself into the furnace.
Gan Jiang gathered 300 boys and girls to pull the bellows. At last, the metal was melted and two swords were forged.
One of the swords is *yang* and the other *yin*. The *yang* sword is called Gan Jiang and the *yin*, Mo Ye.

Bronze

China is among the earliest civilisations to cast bronzeware. Around 2,500 BC, bronze-casting technologies witnessed a rapid rise in China. Being an alloy of copper, tin and lead, bronze has a high strength and a low melting point. During the Shang Dynasty, bronzeware was used mainly for sacrificial and ceremonial purposes. Bronze was also used for making musical instruments and weapons. The designs on the unearthed bronzeware, such as the large set-bells in the tomb of Marquis Yi of the State of Zeng and the sword of King Goujian of the State of Yue, remain clear and brilliant, demonstrating the superb bronze-casting techniques in ancient China. Cast during the Shang Dynasty, the Simuwu Cauldron is the largest and also the heaviest ancient bronzeware in the world.

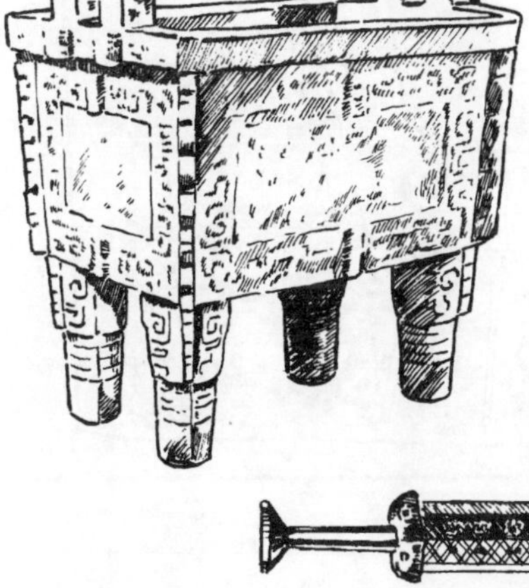

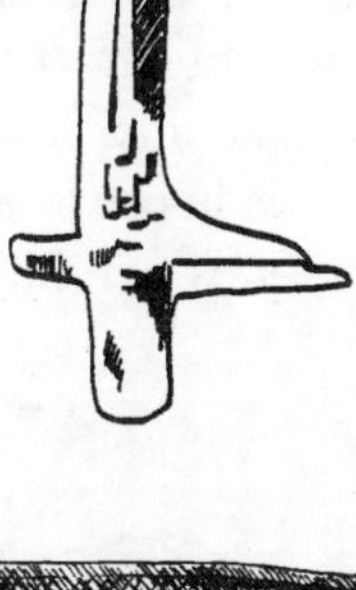

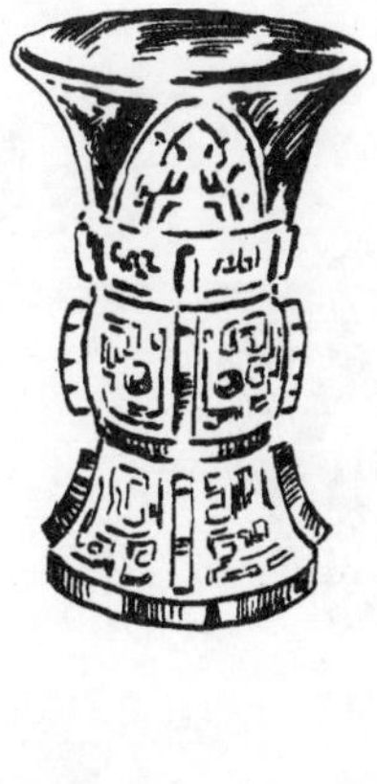

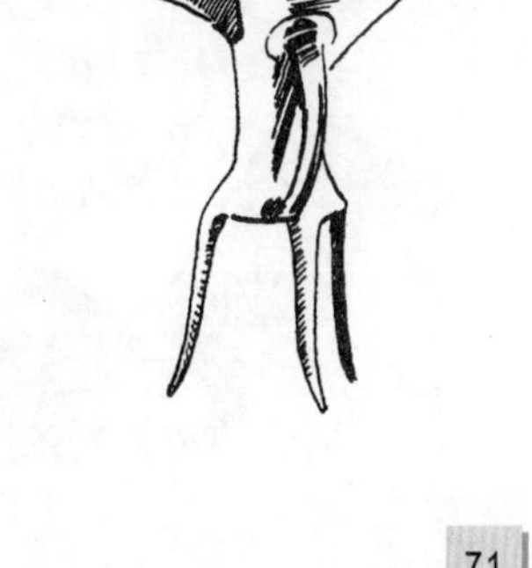

Coal Mining

China is one of the first civilisations in the world to mine and exploit coal on a wide scale for heating, ceramics and metalworking. Marco Polo was surprised to see coal being used in China in the Yuan Dynasty during his tour. In his *Travels* he mentioned that "all over the country of Cathay there is a kind of black stone ... which ... burns like firewood. If you supply the fire with them at night, and see that they are well-kindled, you will find them still alight in the morning." By that time, coal had already been in use as a fuel in China for over 1,000 years.

Development of Agriculture

China is a large agriculture-oriented country which has developed a great variety of farming technologies and tools.

Agricultural Implements

Ploughs

The plough with a curved moldboard appeared in the Tang Dynasty, making it easier to plough the field and save on the animals' power. Thus, cultivating efficiency was increased. The plough was brought to Europe in the 17th century by a Dutch sailor and considerably promoted agriculture there.

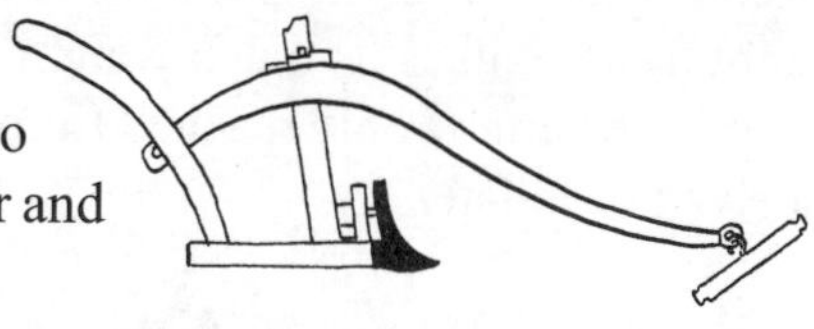

Multi-tube Seed Drills

In ancient times, farmers scattered seeds with their hands when sowing. As a result, the seedlings ran wild in the field, which not only reduced the yields but also made it difficult to harvest the crops. In the second century, the seed drill was invented in China and manual seeding quickly became obsolete. A seed drill only needs a man and an animal to inject the seeds along multiple straight rows at the same time, resulting in increased efficiency and less wastage of seed. This invention did not reach Europe until 1,700 years later.

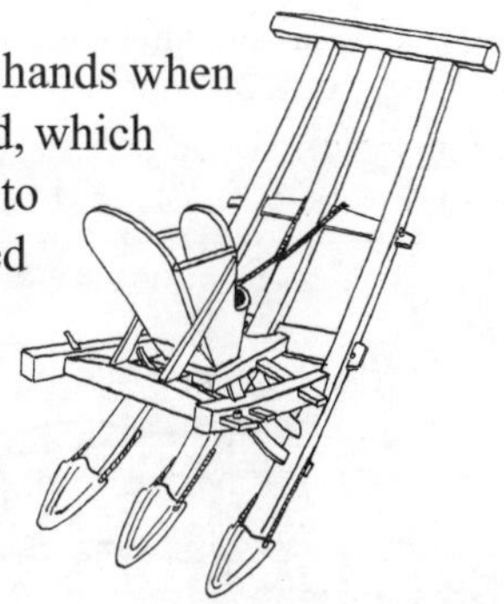

Square-Pallet Chain-Pump (Dragon Backbone Machine)

This water-raising device first appeared in the first century AD and can channel a large volume of water from a low place to a high one, irrigate fields, and drain or supply water. Later, the device was brought to other places in the world and was much welcomed by farmers everywhere. Even today, it is still used in some villages.

Efficient Harnesses

Throughout the world, the ox was first used as a draft animal to plough fields and pull loads, but as time passed, it was replaced by the horse, which is faster and has greater endurance.

However, the harness used for oxen was not suitable for horses because of their different body shapes, and for a long time, horses were unable to work properly because the throat-and-girth harness choked them.

Around 300 BC, the Chinese began to use a harness with a breast strap known as a trace harness, and another century after that, a padded collar harness as well. These new harnesses did not restrict breathing, and the horses were able to pull three times as much weight as they could before with the old harness. The trace harness only reached Europe in the 6th century and the collar harness arrived even later.

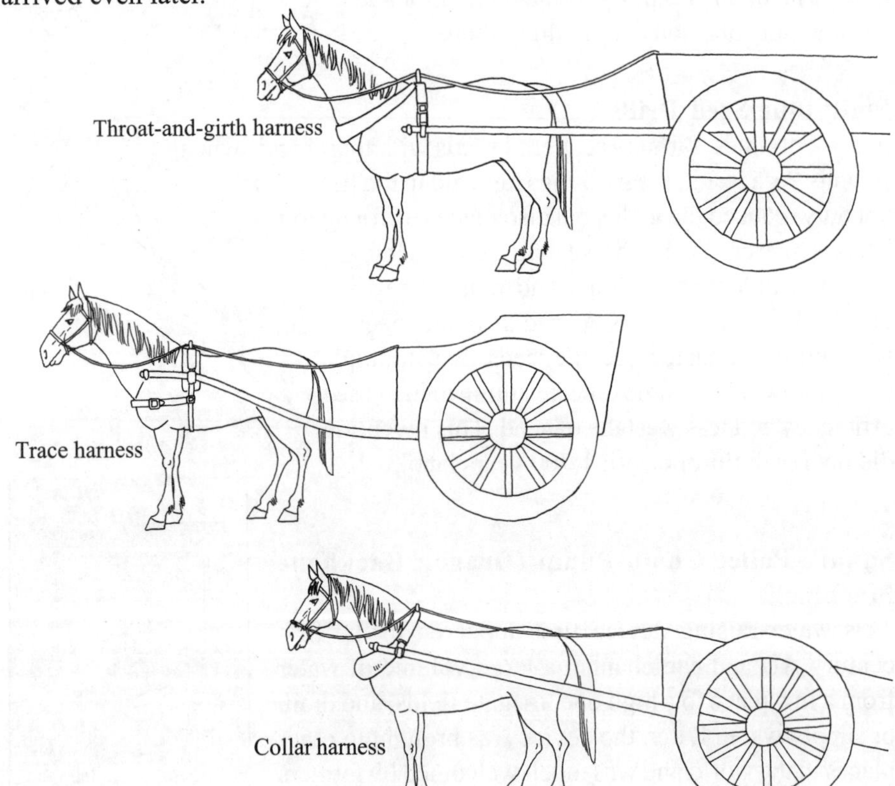

Agriculture

Agriculture has a long history in China. Wet-field rice was cultivated as early as 7,000 years ago, and by the time of the Shang and Zhou Dynasties, most of the main grain crops we have today, such as rice, millet, glutinous millet, wheat, soybean, sorghum and sesame had already been domesticated, as well as vegetables. Gourds were discovered in the ruins of Hemudu dating back to 7,000 years ago, while cabbage and mustard seeds were found in earthen jars excavated from the ruins of Banpo.

Grafting

The technique of grafting one plant onto another was already widely applied in China in the 6th century to improve the shape, colour and quality of fruits and flowers.

Discovery of the Grafting Technique

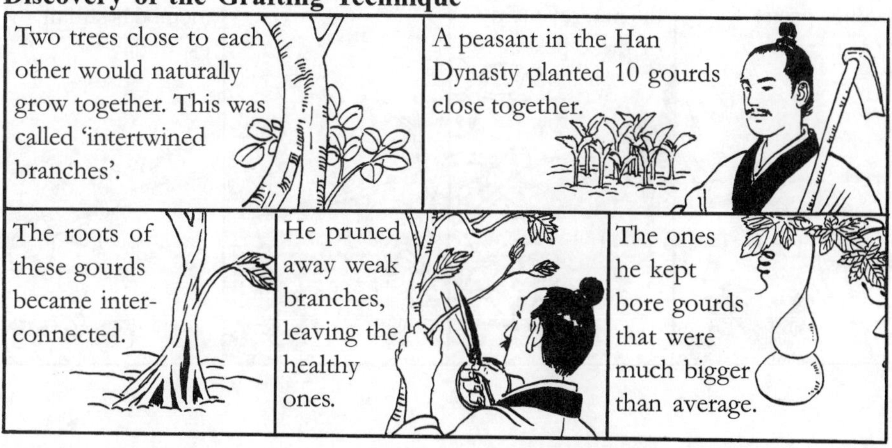

Soilless Culture

The roots and stalks of herbs can be intertwined to form a thick rootstock "cushion". During the Jin Dynasty, people began to grow rice and vegetables on such cushions, which were also known as 'floating fields'. This is actually an early version of soilless culture.

Greenhouse Culture

The Chinese people began to practice greenhouse culture to cultivate vegetables over 2,000 years ago, about 1,000 years earlier than the Europeans. It is said that during the reign of Emperor Qin Shihuang, many hot springs were discovered around the region of Mount Li, and people took advantage of them by building greenhouses to cultivate fruits and vegetables. The imperial palace of the Tang Dynasty set up a special greenhouse that was heated by fire to grow vegetables. During the Ming Dynasty, many wealthy families built greenhouses so they could have vegetables during the colder seasons. With the greenhouse technique, people could get fresh fruits and vegetables throughout the year.

Biological Pest Control

About 1,700 years ago, the Chinese people had already known how to prevent pests with their natural enemies. In the Western world, this trick was not known until the 20th century.

THE TEXTILE INDUSTRY

It is generally known that China is the homeland of silk. People of China learnt to raise silkworms and weave silk into fabrics long ago.

Raising Silkworms and Weaving Fabrics

Silkworms are the caterpillars of the silk moth. They produce very fine and strong silk fibres when spinning cocoons to prepare for the pupa stage.

Silkworms feed on mulberry leaves. Many mulberry trees are needed to raise silkworms.

The silk fibres are spun into thread and woven into beautiful fabrics

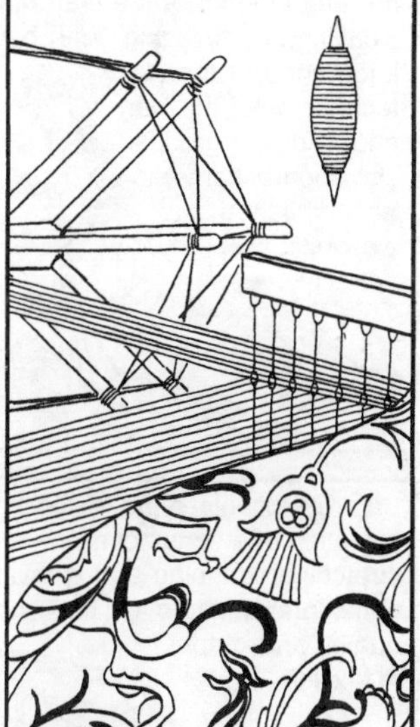

The Chinese people had already invented a simple weaving loom as early as the Neolithic Period.

The Story of Huang Daopo

ENGINEERING MARVELS

The Chinese have many awe-inspiring engineering accomplishments to be proud of. These showcase the skill and determination of the Chinese people. These great projects are also milestones in Chinese technology and science.

THE GREAT WALL OF CHINA

The Great Wall of China is one of the world's greatest engineering achievements. During the Warring States Period, the dukes and princes built fortresses and walls along their borders. When Emperor Qin Shihuang unified China, 300,000 men were employed over a span of more than 10 years to connect, reinforce and expand the northern walls of Yan, Zhao and Qin to build the meandering 5,000 km long Great Wall of China that stretched westwards to Lintiao and eastwards to Liaodong.

From post-Qin to the Ming Dynasty, the Great Wall underwent constant repairs and construction. During the Ming Dynasty, the Great Wall was extended from the western Jiagu Pass to the eastern Shanhai Pass, a total length of over 6,700 km.

It is said the estimation of the building materials required for the building of the Jiagu Pass was so accurate that only one brick was left after the completion of the wall.

Tenon and Mortise

7,000 years ago, the forefathers of the Chinese people had already started using wood to build houses. At that time, the tenon and mortise method was already used to join pieces of wood.

The mortise (eye) is the part with a hollow. The tenon is the protruding part that is inserted into the mortise.

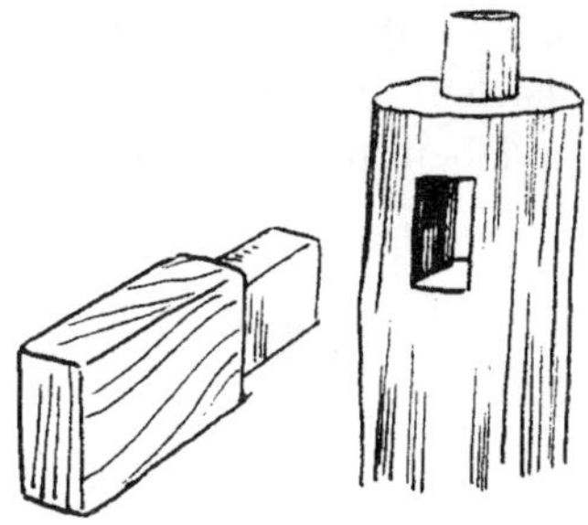

When the tenon and mortise meet, the wooden pieces will be firmly joined even without the use of nails. This is one of the foremost inventions in Chinese architecture.

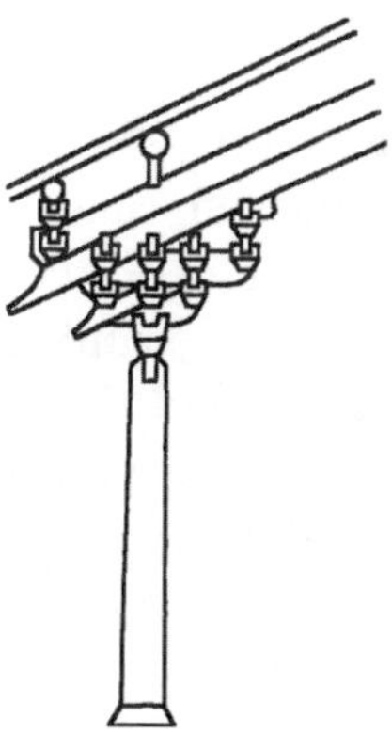

Dovetail

Dovetails connect and join the pillars to the roof. They buttress the beams and pillars, making the building more solid and stable. Dovetails raise the ends of a roof, enhancing the appearance of a building while improving the ventilation as well. Dovetails first appeared about 2,000 years ago.

Decoration

Ancient buildings had auspicious figures placed on their roofs to avert disasters. They were also symbols of peace and safety. The grotesque ceramic animals placed at either end of a roof ridge serve as ornaments and also stabilise the rooftop. The tiles and drip tiles shelter the building from rain and protect the walls.

The Evolution of Human Dwellings

Caves

In the beginning, before humans knew anything about the construction of buildings, they found their dwellings in nature.

Tree Nests

The cavemen went on to build houses on trees with tree trunks, branches and dried leaves. The result resembled a bird's nest, hence the name 'tree nests'.

Houses on Stilts

As the tree houses were built on trees, they could only live where there were trees. The ingenious human race then came up with using wooden stilts to replace tree trunks. So they began building houses raised on stilts to live in.

Den Dwellings

The tree house is similar to a bird's nest. But there were others who chose to burrow underground like rodents and live in holes below the ground!

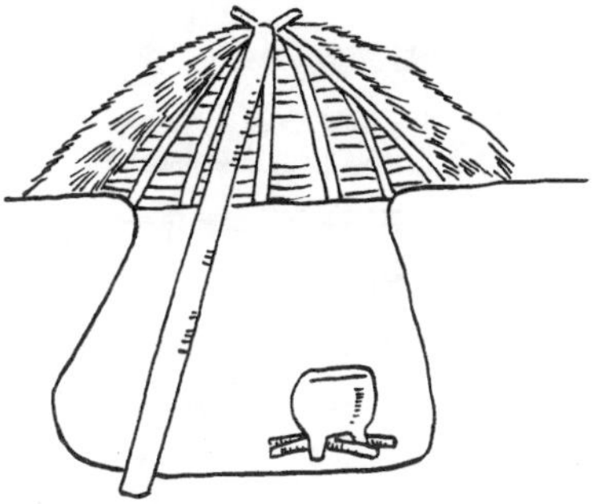

When it rains, won't the den flood? So the den-dwellers erected a canopy above the hole to keep the rain out.

Half-den Dwellings

Going in and out of a den or an underground hole is troublesome. So humans stopped burrowing too deep underground and began to build roofs further above the ground.

Surface Houses

To support the rooftop and provide shelter from wind and rain, walls were built around the house. Thus, the den dwelling evolved into the surface house. In the early days, the entrance to the house was in the roof!

Types of Buildings

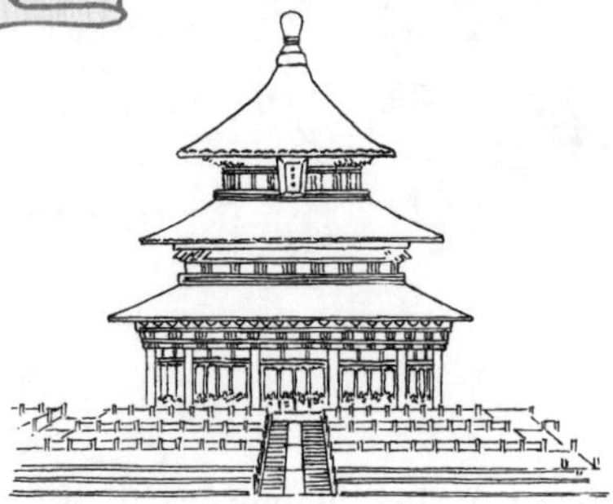

Temples
These are built for prayer ceremonies and rituals. The famous Temple of Heaven in Beijing was where prayers to heaven and prayers for a good harvest were conducted during the Ming and Qing dynasties.

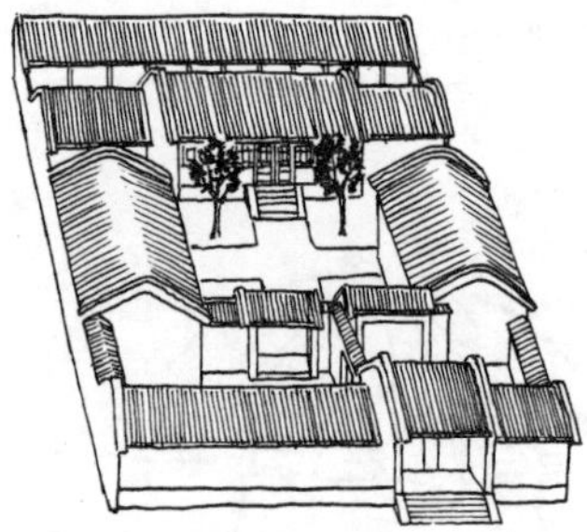

Residential Structures
These are housing for the ordinary people. Each area has its own distinctive architectural style. The *siheyuan* has houses around a square courtyard, with north, south, east and west wings. The area is enclosed by a wall to form a square compound.

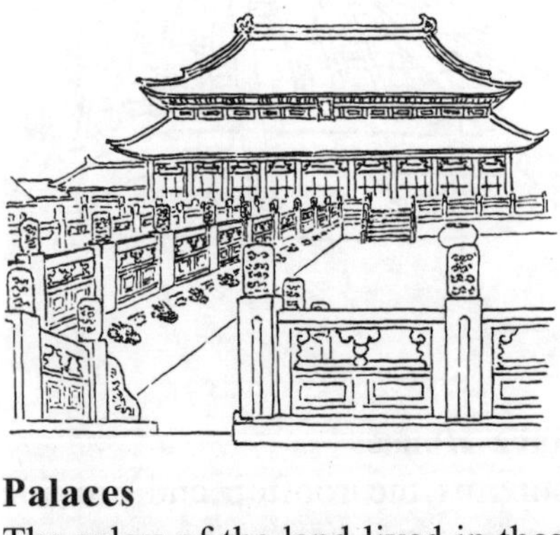

Palaces
The rulers of the land lived in these grand residential buildings during ancient times. The world-renowned Forbidden City (the Imperial Palace) contains scores of palaces and courtyards.

Gardens and Parks
Ancient Chinese gardens and parks are on a large scale, with mountains, streams, flowers, trees, towers and pavilions laid out in picturesque compositions. One of the most famous parks is the Suzhou Garden.

Pavilions

These are small shelters built by the road or in a garden to offer a resting place for travellers.

Multi-storey Buildings

These may be square, hexagonal or octagonal in shape. All sides have open windows and most are built on high ground.

Pagodas

These towers, usually built for religious or memorial purposes, range in height from five to 13 storeys.

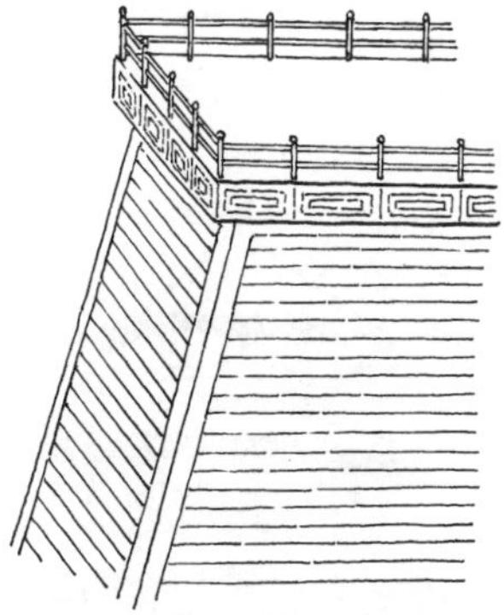

Platforms

These are tall buildings with a flat roof to provide a high vantage point.

Terraced Pavilions

Often built beside water, these structures provide a place for recreation and relaxation. They are also known as waterside pavilions.

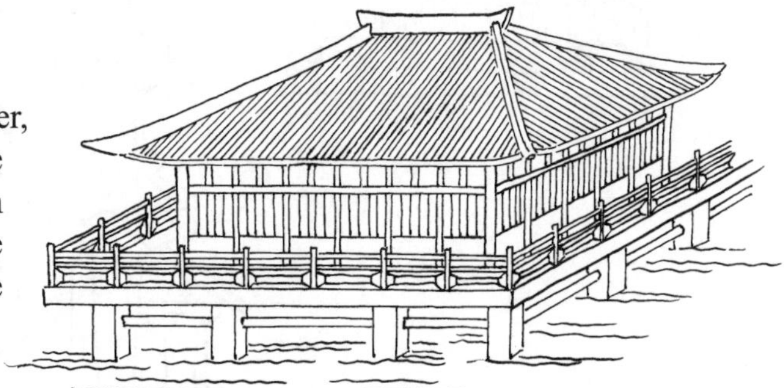

China's Architectural Mosts

The world's tallest ancient wooden pagoda still standing is in Shanxi, China. The Wooden Pagoda in Ying County has stood for over 900 years, weathering wind and storms and enduring earthquakes through the years, making it an architectural miracle.

The Wood Pagoda was built only with dovetails and the tenon and mortise method. Not a single nail or rivet was used. Over 50 types of dovetails were used in the construction. It is a masterpiece of meticulous planning and design, elegant and imposing.

The World's Largest Palace

The Forbidden City is the largest imperial palace extant in the world. Built in 1420, the palace housed the imperial families of the Ming and Qing dynasties. It has a history of over 500 years.

The grounds contain scores of large and small courtyards, over 70 palaces and more than 9,000 houses. Each block of buildings runs from south to north. The Forbidden City is protected by a wall more than 10 metres tall and a moat more than 50 metres wide.

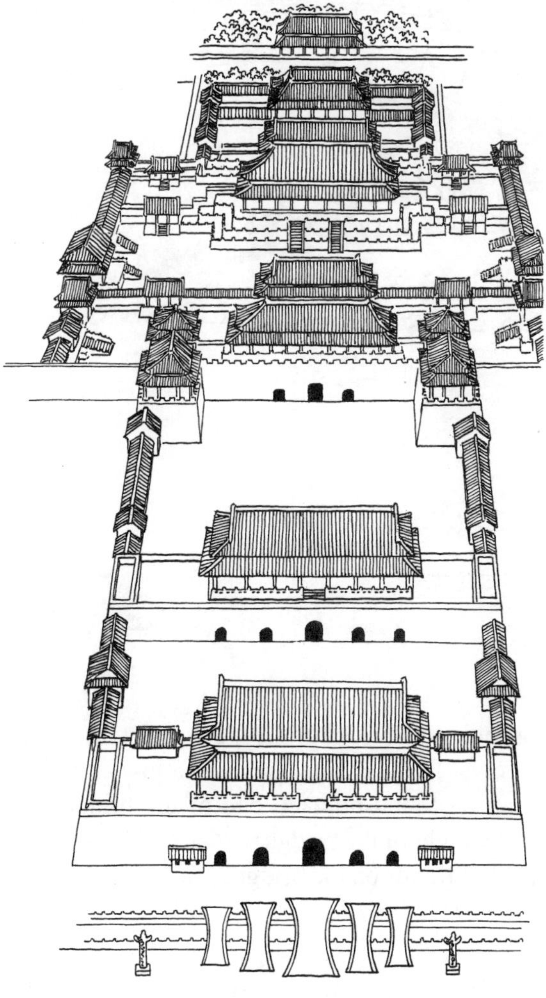

The World's Oldest Stone Arch Bridge

The Anji Bridge (commonly known as the Zhaozhou Bridge) in Zhao County, Hebei was built between AD 590–608. For more than 1,000 years, despite numerous floods, earthquakes and long usage, the bridge still remains solid and strong. The bridge structure is known as the 'open shoulder'. This bridge design would only appear in Europe more than 700 years later.

The Anji Bridge was built with bricks of the green and white limestone available locally, arranged in rows. It has a wide span and a low arch. Additionally, two small arches each can be found on each end of the bridge arch. This design feature distributes the weight of the bridge and also allows water to flow freely, minimising the impact of floods on the bridge.

Other Types of Bridges

Single-plank Bridges

Single-plank bridges made their first appearance about 6,000–7,000 years ago. They were designed to be mobile rather than fixed in place. During the day, such bridges would be let down to allow people to cross, while at night, to prevent wild animals from encroaching, the bridges would be put away.

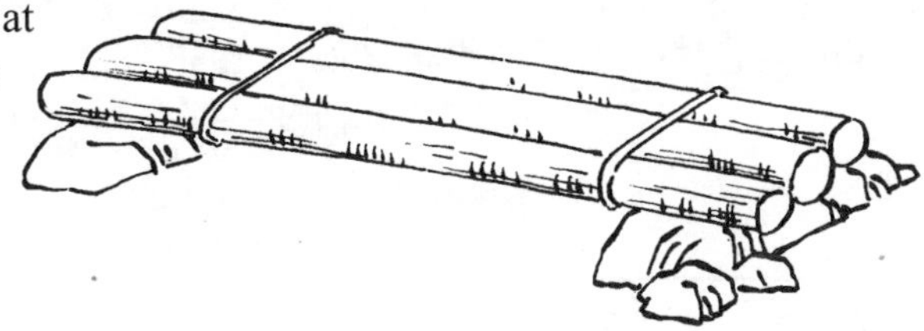

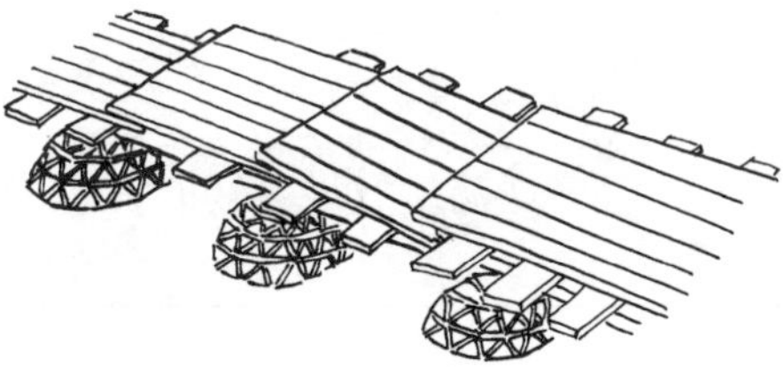

Floating bridges

Wooden boards are placed across two objects floating side by side to serve as portable bridges. Objects used include boats, rafts, bamboo pallets, wooden pallets and even wooden wheels. In ancient times, these bridges were often used for military troop movements.

Beam Bridges

Beam bridges comprise a primary structure of wood or stone that extends across standing pillars.

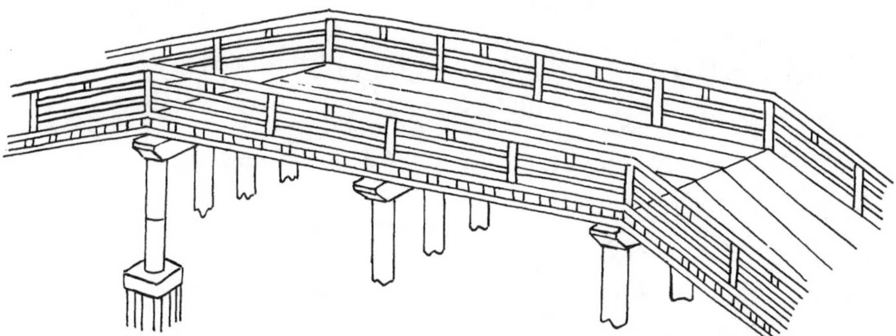

Suspension Bridges

These bridges use thick cords to link two opposite ends. The cables may be made of bamboo, metal or rattan. Suspension bridges are normally found in mountainous areas with rapids that are unsuited to building bridge pillars.

How Suspension Bridges Are Built

1. Attach a thick cord to the arrow and shoot it across.

2. Attach the thick cord to a spear and throw it over to the opposite side.

Fengshui

Fengshui is a study of the geography of a residential area, arrangement of furniture and items, as well as the location of a burial ground to decide if they will promote well-being or invite disaster.

Fengshui combines the study of geography and astronomy. Through a better understanding of one's surroundings, changes can be effected to create a living environment that will benefit the resident, thus producing the ideal home. It is important to take a holistic approach to the environment. As people interact with their environment (both nature and home), changing the surroundings will also affect them in turn.

In *fengshui*, mountains represent the pulse of a dragon. This is actually the study of modern geology, geography, mountain ranges, water flow and rock formations. Water quality is determined by its mineral content, which may benefit or harm a person. Traditional *fengshui* study involves the tasting of the earth and water. If the water tastes sweet, the land is good. If the water is bitter, it is an ominous land.

Qi is considered the source of life. Without it, there will be no life. Hence *fengshui* advocates building on a land that has qi, to give the dweller good health and longevity. The entrance to the house is the pathway of qi. If the entrance is surrounded by water and it meanders, there is qi. If the main door is in a stuffy area, there is no qi.

Chinese Gardens

Beautifully landscaped Chinese gardens are world-renowned. They may be classified into the northern imperial gardens and the Jiangnan private gardens.

Northern Imperial Gardens
These sprawling grounds were where members of the imperial family entertained, hunted, relaxed and hosted official functions. Famous ones include the Summer Palace and Chengde Summer Villa.

Private Gardens in Jiangnan
Most of them are exquisite, small and boast of scenic spots and intricate architecture. Examples are the famous Zhuozheng Garden in Suzhou, Liu Yuan and Lunlang Pavilion.

Features of a Chinese Garden

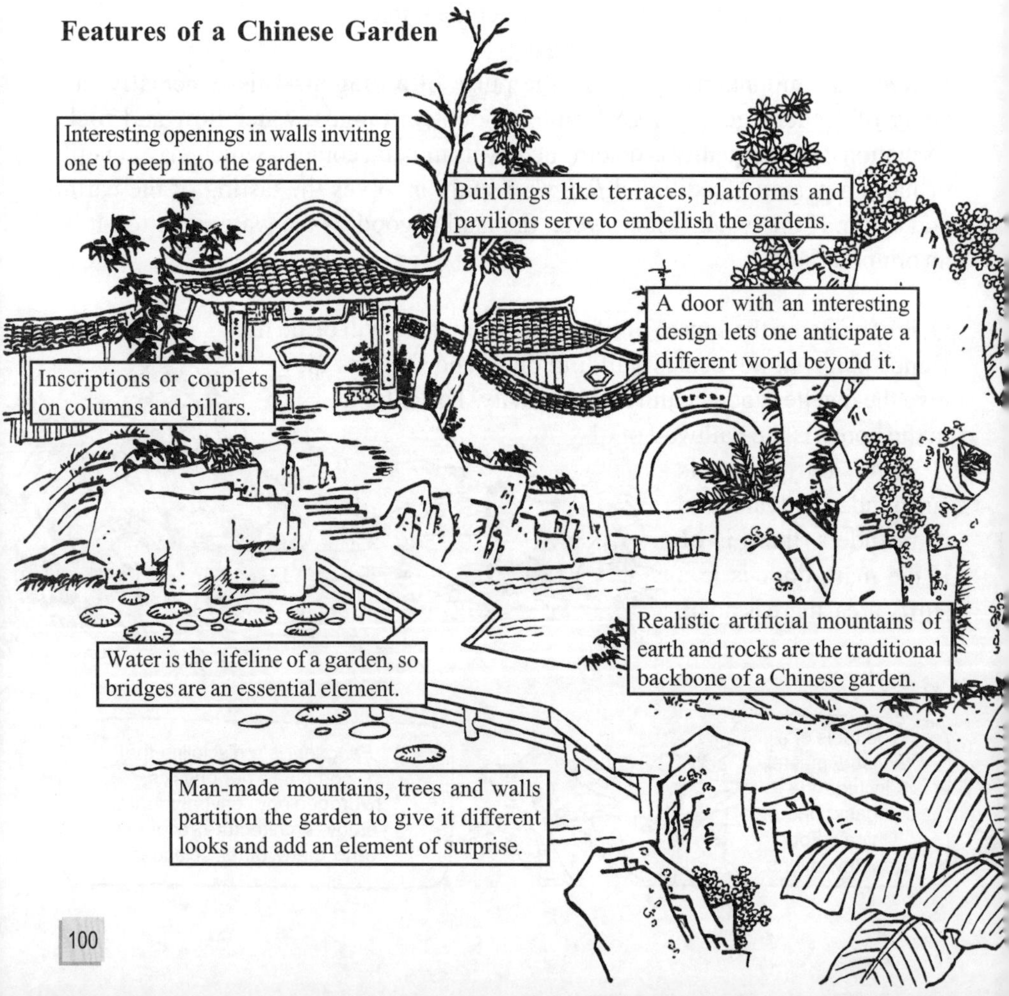

Other Inventions

While the four great inventions of China are well-known, many other familiar, everyday objects — the saw, the matchstick, the mirror and even the rocket — are also Chinese inventions.

The Invention of the Rocket

To this day, rockets remain a vital component in space travel. They carry man-made satellites, guided missiles and space shuttles into space. Feng Jisheng and Yue Yi, two military officers during the Northern Song Dynasty, invented the first rocket.

The body comprises an arrow and a gunpowder tube. At the end of the tube is a fuse.

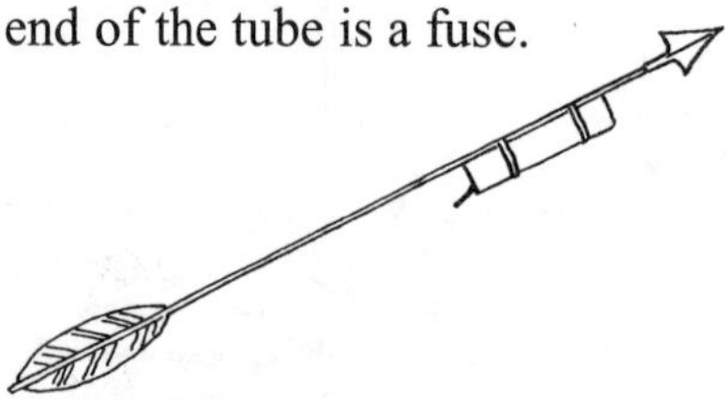

After the fuse is lit, the burning gunpowder will propel the rocket forward.

The manufacture of rockets became more sophisticated as time went by. Some of them could cover distances ranging from 457 to 1,052 metres.

The Chinese people also came up with a stabiliser to control the rocket's flight path. This device was similar in appearance to a pair of bird's wings, so the rocket seemed to be attached to a bird. In later years, the Chinese invented many other types of rockets, stamping their mark on the field of rocketry.

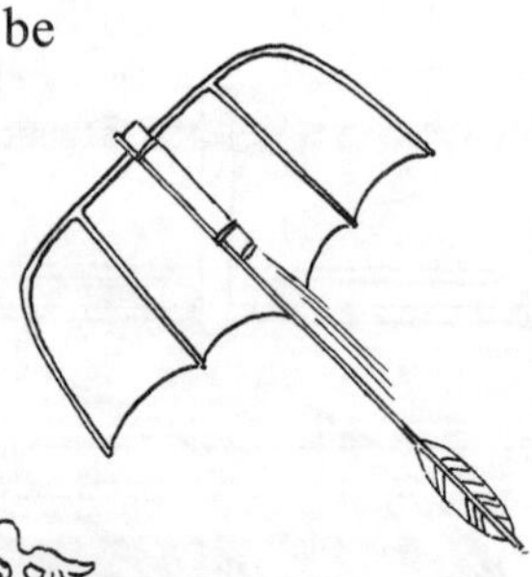

Chinese rocketry later spread to Europe along with gunpowder. The first European rocket weaponry first appeared towards the end of the 14th century. The invention of liquid fuel combustion in the 20th century ushered in a new era in the human race's exploration of space.

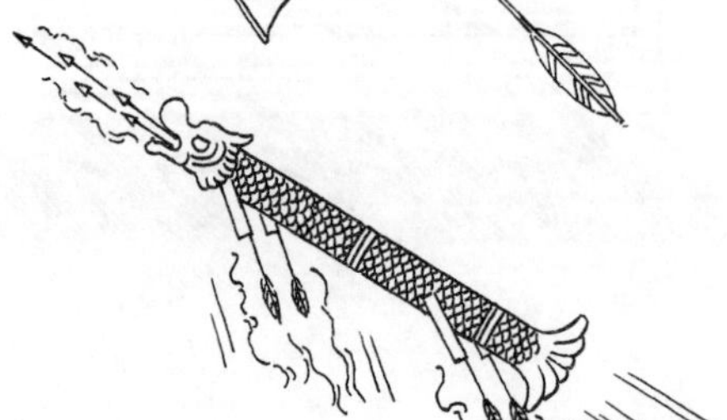

Wan Hu's Attempt at Rocket-powered Flight

When the deadline arrived, he had the flying bird ready.
This flying bird will be propelled by the rockets behind me. Let's see how far it can fly.
That's Wan Hu in his flying bird!
With a loud boom, the flying bird flew into the sky, leaving a trail of fire in its wake.
Wow! It's really flying!
Amidst the cheers, the rockets suddenly stopped burning...
... and the flying bird plunged into the valley.
Wan Hu crashed to his death. Although his experiment failed, the International Astronomy Association christened a lunar crater on the moon Wan Hu in honour of his heroic attempt.

The Story of the Saw

Lu Ban is China's foremost artisan in history. He was also a prolific inventor who created the saw, plane, artisan's square, drill, shovel, artisan's ink marker, millstone and scaling ladder, among others. He once sculpted a bird out of bamboo which managed to fly in the sky for three days and three nights. This prototype of the kite marked one of mankind's earliest attempts at flight.

The Invention of the Hot Air Balloon

200 years ago, the French discovered the hot air balloon. It was later used in sending mail, transporting bombs and espionage during war times. However, before the invention of the hot air balloon, the Chinese people had already used similar contraptions.

The ancient Chinese understood the lifting property of hot air. In 2 BC, a flammable substance was placed in empty egg shells. Upon ignition, the egg shells shot into the sky. This was the world's first scientific experiment testing the principle of hot air.

During the Three Kingdoms Period, the famous military strategist Zhuge Liang invented the Kongming Lantern. Made of paper, it functions on the premise that hot air is lighter and will thus rise. There is no opening on the top of the lantern. When the lantern is lit on the inside, hot air soon fills the interior, causing the lantern to rise into the sky.

The Story of the Kongming Lantern
During the Three Kingdoms Period, Zhuge Liang, the Prime Minister of the state of Shu, led the army against the states of Wei and Wu.
This terrain is easy to defend. But we will have problems transmitting military updates as well as co-ordinating our battle efforts.
How do I send my orders to the army?
Cut some thin strips of bamboo and make large paper lanterns with them. Leave an opening only at the base of the lantern.
Yes, Prime Minister!
We've put slow-burning resin in the centre of the lanterns as you ordered.
It works just like I hoped.
Excellent. Now we can use these lanterns to send orders and information quickly.
The Shu army worked out a code based on the number of lanterns in the sky, the height of the flight and markings on the lanterns.
As Zhuge Liang was also known as Kong Ming, the lantern was called the Kongming Lantern. It works based on the same principles as the hot air balloon.

Ancient Robots

Chinese artisans were said to have created life-like puppets as early as 2,000 years ago. These mechanical robots could move, dance, see and speak. They were mostly used for entertainment.

The Tang Dynasty saw many varieties of exquisite automated ornaments. Ma Daifeng, a top artisan of Jiangsu, created an automated dressing table for the Empress. When the Empress wished to groom herself, all she had to do was open the lid to the dressing table and a wooden female figurine would appear with a towel and hairbrush. Figurines holding rouge, powder and other cosmetics would then appear in line to wait on the Empress.

Ma Daifeng also created the 'wine mountain'. It was a model of a mountain summit with an automated figure in a loft. This automaton would walk out to urge customers to drink wine. If a person took a long time to finish the wine, the robot would appear again to urge the person to drink it up.

King Zhou Mu's Puppet

Legend has it that as King Zhou Mu was on his way back after touring the western region, a top artisan named Yan Shi was given to him.

The Earliest Slide Projector — The Running Horse Lantern

The running horse lantern, also known as the divine lantern, consisted of a series of drawings on fine paper. These drawings would move in sequence, giving the illusion that the pictures were actually moving.

The running horse lantern used heat or manual labour to turn the pictures. The pictures could be projected onto the wall. What we know as projector slides have their origins in the Chinese running horse lantern.

The Matchstick

A palace maid of the ancient Northern Qi Dynasty in China invented the matchstick. It wasn't until 1,000 years later that the matchstick was first seen in Europe. In 1830, the French and Germans made improvements to the humble matchstick. They mixed yellow phosphorous, sulphur and potassium chloride together to create the matchstick we know today.

The Invention of the Matchstick

The Fishing Reel

The fishing reel was discovered in AD 3. It is said that the fishing reel was inspired by the ballista, a weapon invented by the Mozi School during the Warring States Period. The fishing reel only made its appearance in Europe in 1651, more than 1,300 years later.

The Pinhole Projection Experiment

Mozi, a philosopher from the Warring States Period, used to be a craftsman. He also studied optics, mechanics, mathematics and other subjects. Mozi conducted the world's first pinhole projection experiment, proving that light travels in straight lines before the equivalent Greek optical experiment.

Make a small hole in the wall that faces the sunlight. An image of the person standing outdoors facing the hole will be projected upside down on the opposite wall in the darkened house.

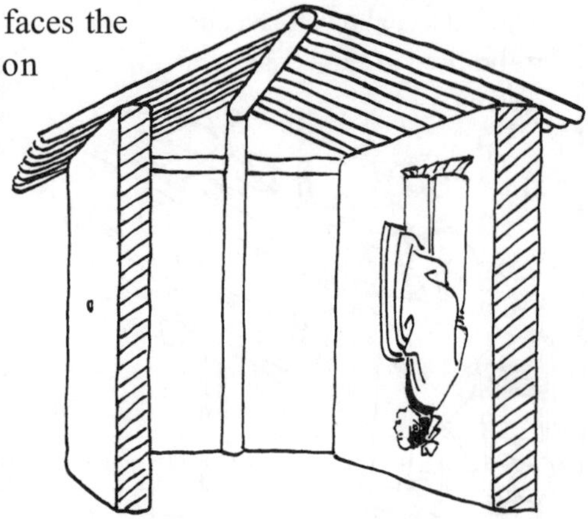

Mozi explained that light rays travel in straight lines through the hole like arrows. The person's head block the light rays from above and result in the image of the head projected below; the person's feet block the light rays from below, resulting in the image of the feet being projected above. The end result is an upside-down image of the person.

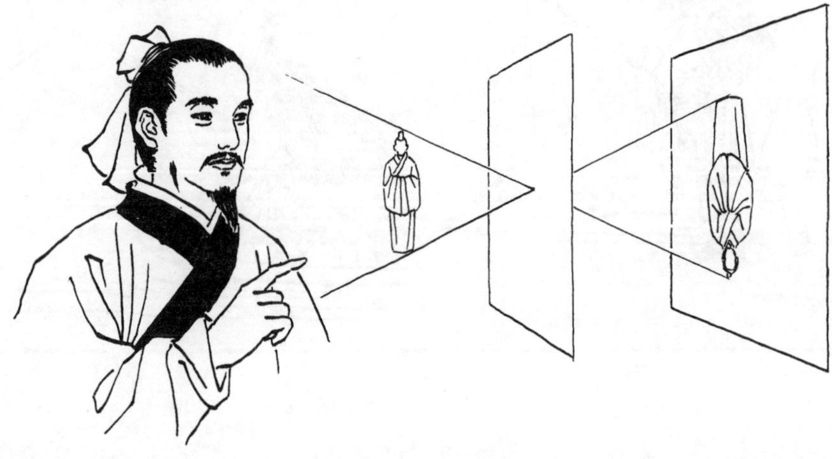

Make Your Own Pinhole Camera

1. Obtain a cardboard box. Remove the top and paint the interior of the box black.

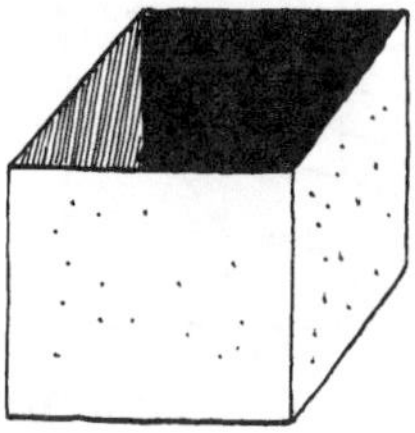

2. Cover the exposed top with a piece of tracing paper.

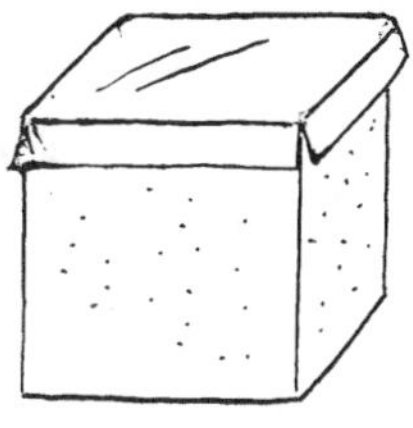

3. Use a needle to make a tiny hole in middle of the side of the box opposite the tracing paper.

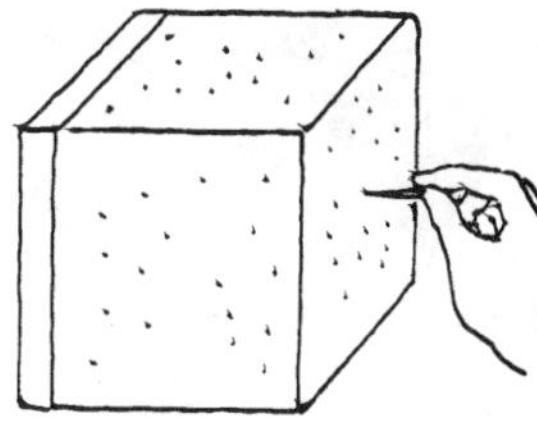

4. Take the completed box outdoors. Place the side with the tracing paper in front of you.

5. Cover your head and the box with a piece of cloth.

6. Aim the hole at any object. Its image will appear on the piece of tracing paper.

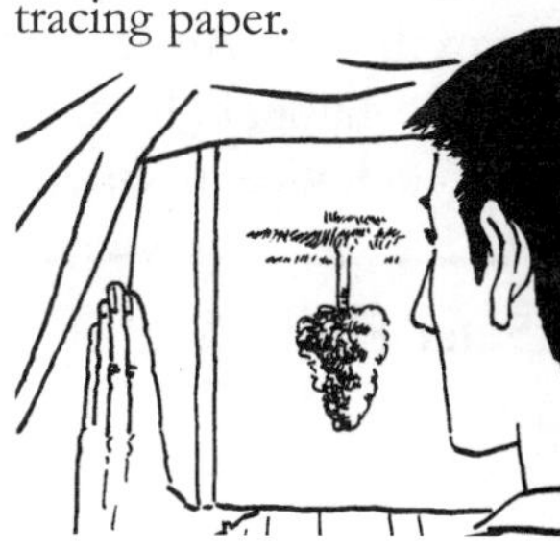

The Principle Behind Pinhole Projection

Light rays reflected from the object pass through the pinhole and fall on the tracing paper. The rays cross one another as they pass through the pinhole, resulting in an inverted image.

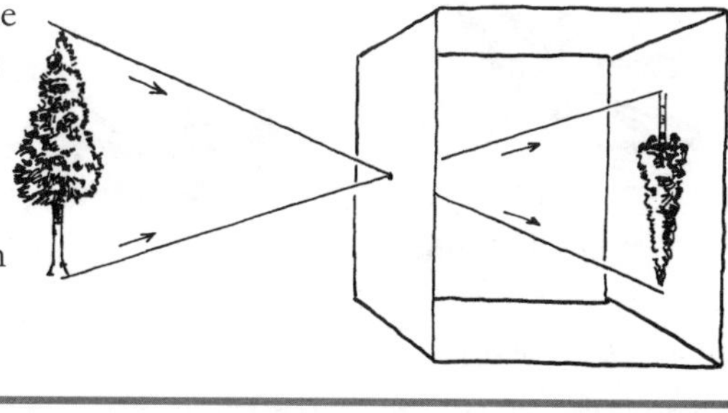

The Mirror

Light travels in straight lines. But when the path of light rays gets blocked by an object, the rays will be reflected. This principle of light reflection was used by the Chinese 3,000 years ago to create the bronze mirror.

The Principle Behind Mirrors

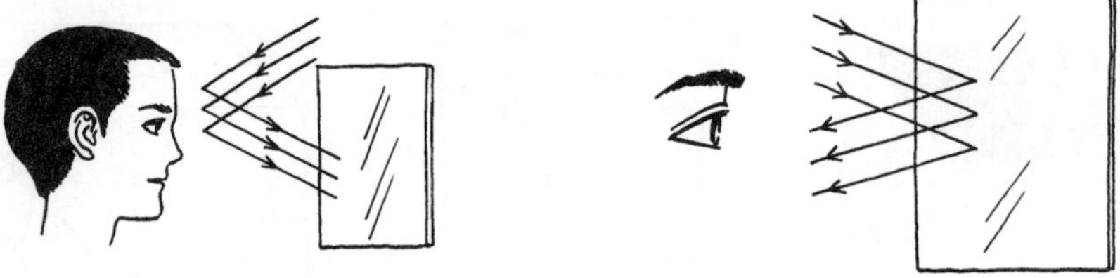

When light rays hit an object, they will be reflected onto a mirror.

Light rays bounce off the mirror to enter our eyes, so we see the object as a reflected image.

The flat mirror also gave rise to the discovery of the amazing concave and convex mirrors. Ancient Chinese manufactured the solar flint. It used the concave mirror to gather light energy from the sun in order to ignite. It was the first solar-powered device invented.

Solar Flint

Bronze was fashioned into a concave mirror to gather light energy.

It was placed out in the sun at noon.

The moxa placed on it soon caught fire.

During ancient times, makers of the convex mirror were also very familiar with its function. If the mirror was small, they would make its surface more convex so the entire face could be seen in it.

Concave and Convex Mirrors

1. Take a soup spoon with a shiny surface.

2. Look into your own reflection on the convex surface. Does your face seem to be bigger?

3. Next, turn the spoon around and look into the concave surface. What difference do you notice?

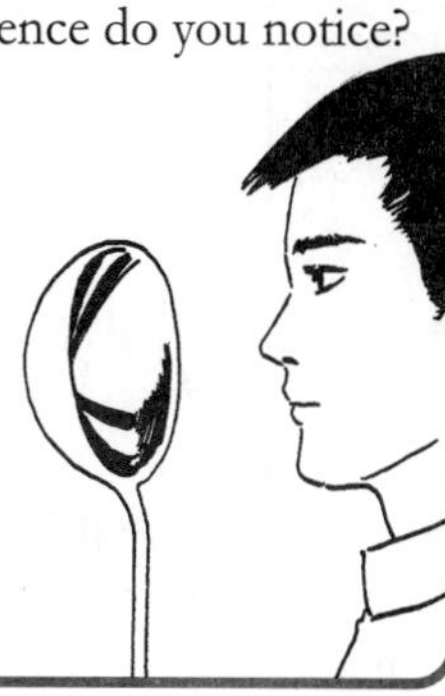

Applying the principle of the flat mirror, China produced the world's first periscope in 2 BC. A book during the early Han Dynasty, entitled *The Ten Thousand Infallible Arts of the Prince of Huainan* (*Huai Nan Wan Bi Shu*), wrote: "Hang the mirror up high and you will be able to survey the surroundings above water while you are under water." Periscopes used by the present-day navy work by this same principle.

Make Your Own Periscope

1. Find a long and narrow cardboard box. Make parallel diagonal cuts on both ends of the box.

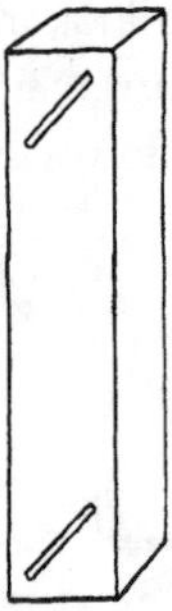

2. Cut openings at the two ends where shown.

3. Slide a mirror into the diagonal cut, with the mirror facing the opening. Secure with adhesive tape.

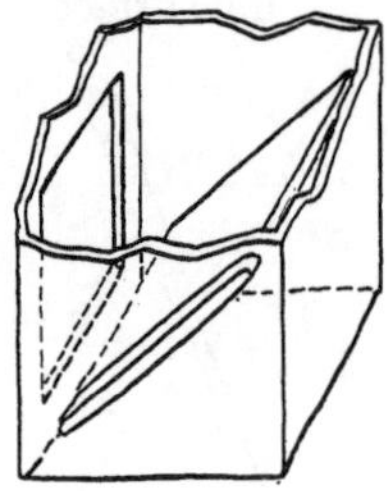

4. Repeat with the other end. Check that both mirrors are firmly fixed in place.

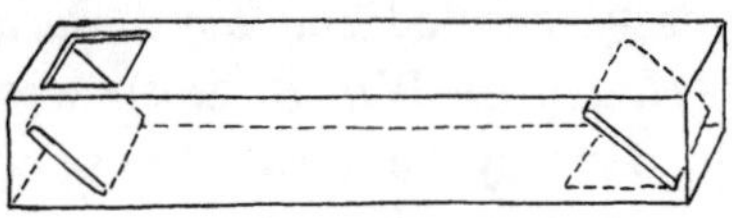

5. The periscope is now ready to be used. Use it to look around corners, over a high wall or through a window.

How the Periscope Works

The light bouncing off the objects from the outside hits the top mirror.The top mirror then reflects the light rays onto the bottom mirror, and from there into our eyes. That is how we are able to see the objects outside the window.

The Horseshoe and Stirrup

For millennia, horses were used to transport goods and people. Wars were fought with soldiers on horseback and in chariots. The invention of the horseshoe and stirrup played a key role in the domestication of horses.

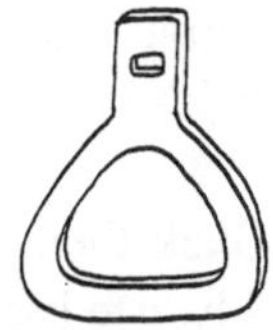

China began using horseshoes over 2,000 years ago while simple leather horseshoes were still being used in Europe. When the horseshoe was invented, it was quickly copied by other nations and soon became widespread.

Stirrups were invented around AD 300, and allowed riders greater stability, increasing the range of mounted manoeuvres possible. They were later introduced to Europe where they ushered in a new era of mounted warriors during the Middle Ages.

An Interesting Fact

Turkish horsemen employed the Chinese stirrups to great effect against the Eastern Roman Empire in AD 560. The Romans copied the stirrups used by the Turks and made modifications to their cavalries. With their iron stirrups, the Eastern Roman Empire were then able to effectively defeat their foes.

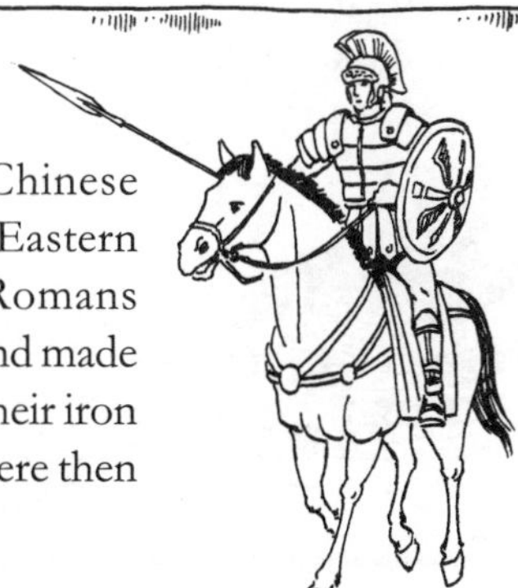

The Rudder

The rudder is a device that controls the direction that a boat travels. Ancient Chinese based their design of the rudder on their study and observation of the fish's tail.

Back then, Chinese ships and boats were the most advanced of their time. Chinese nautical achievements were way ahead of the competition in other parts of the world.

▽ **The Invention of the Scull**
The scull is used to propel the boat forward. It is said that Lu Ban invented the scull.

△ **The Invention of the Sail**
The Chinese put sails on their boats 3,000 years ago. The sail makes use of wind to move the boat.

GREAT ANCIENT SCIENTISTS

The history of China has seen many great scientists. With their determination and the rigour with which they pursued their studies, they made great contributions in the arena of science and technology which kept them on the cutting edge far ahead of their time.

Zhang Heng

Zhang Heng (AD 78–139) was a multi-talented scientist of the Eastern Han Dynasty. He was also an accomplished literary figure, painter, and even worked as an official overseeing policies on astronomical movements.

While in office, he made great contributions to the study of astronomy and earthquakes. He invented the world's first equatorial armillary sphere and the first seismograph. His book, *Mystical Laws* (*Ling Xian*) is China's first major work on theories of astronomy.

To commemorate Zhang Heng's scientific achievements, the International Astronomical Union named a lunar crater on the moon after him. Asteroid #1802, which was discovered in 1964, is also named in his honour.

Zhang Heng's grandfather Zhang Kan was a founding father of the Eastern Han Dynasty, though the family's fortunes had declined by his father's time.

When Zhang Heng was 10, his father passed away, leaving his grandmother and mother to bring him up. Since childhood, Zhang Heng loved literature and aspired to become a writer.

At 16, he left his hometown in Nanyang for Luoyang — the cultural hub at that time — in search of a master who would take him in.
It took Zhang Heng 10 years to complete the *Er Jing Fu*, the book that earned him a reputation as a great writer.
He also came across the writings of the philosopher Mozi, and Yang Xiong, a writer and philosopher from the Western Han Dynasty. These influenced him to study natural science.
When he was 38, he was put in charge of research on astronomy. He also handled rituals and imperial ceremonies such as prayers, weddings, funerals and others.
Although this isn't a post that is highly sought-after, it will allow me to concentrate on my scientific studies and research.
During his six years in office, he invented the equatorial armillary sphere and wrote a book on astronomy entitled *Ling Xian*.

Zhang Heng was later promoted to be the commander of the chariots, a high post that was coveted by many. But he remained devoted to his research.
In AD 126, Zhang Heng was put back in charge of astronomy and palace rituals and ceremonies.
Most people get promoted. But you get demoted instead.
To each his own. I enjoy what I'm doing. So what does it matter?
During Zhang Heng's lifetime, earthquakes were frequent. These natural disasters struck fear in the people's hearts.
Report any earthquakes to the Capital.
However, news of earthquakes in faraway places took too long to reach the capital.
To solve that problem, Zhang Heng created the world's first seismograph in AD 132.
Now I will finally know when there are earthquakes.
Ah! That means an earthquake has occurred west of Luoyang!

* *A Chinese unit of measurement that equals 0.5 kilometres.*

Zu Chongzhi

Zu Chongzhi (AD 429–500) was a mathematician, astronomer and engineer of the Southern Dynasties. He calculated the value of pi as somewhere between 3.1415926 and 3.1415927. He also established the most advanced calendar of his time, the Daming Calendar, and invented many complex mechanical devices like the water-powered trip hammer and the thousand-*li* boat. In honour of his achievements, Crater Zu Chongzhi on the moon was named after him.

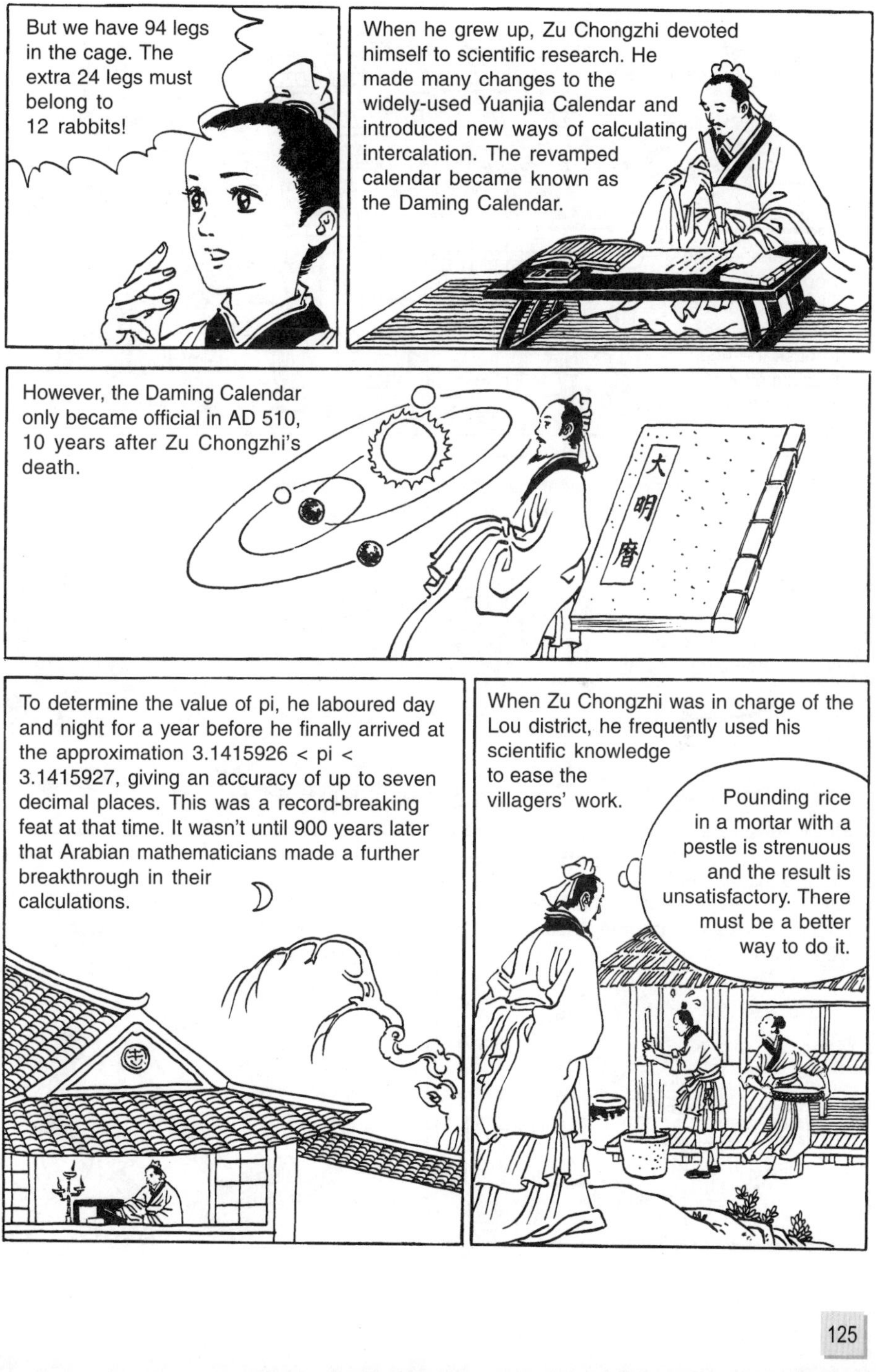
But we have 94 legs in the cage. The extra 24 legs must belong to 12 rabbits!
When he grew up, Zu Chongzhi devoted himself to scientific research. He made many changes to the widely-used Yuanjia Calendar and introduced new ways of calculating intercalation. The revamped calendar became known as the Daming Calendar.
However, the Daming Calendar only became official in AD 510, 10 years after Zu Chongzhi's death.
大明曆
To determine the value of pi, he laboured day and night for a year before he finally arrived at the approximation 3.1415926 < pi < 3.1415927, giving an accuracy of up to seven decimal places. This was a record-breaking feat at that time. It wasn't until 900 years later that Arabian mathematicians made a further breakthrough in their calculations.
When Zu Chongzhi was in charge of the Lou district, he frequently used his scientific knowledge to ease the villagers' work.
Pounding rice in a mortar with a pestle is strenuous and the result is unsatisfactory. There must be a better way to do it.

Zu Chongzhi came up with the water-powered trip hammer and taught the villagers to use it.
This hydraulic mill is amazing! Husking the rice will no longer be such a laborious chore.
The moving water pushes the wooden wheel, which then turns the trip hammer to husk the rice. It saves a lot of human effort.
Zu Chongzhi also designed a boat that was propelled by wheels. It could move very fast and everyone called it the thousand-*li* boat.
Wow, watch it go! It's really a thousand-*li* boat!
He also made improvements to the south-pointing carriage and the water clock, and re-created Zhuge Liang's legendary wooden ox.
Zu Chongzhi's talents extended to music and literature, and he even wrote a book of ghost stories called *Shu Yi Ji* (Accounts of the Strange and Bizarre).
述異記

Shen Kuo

Shen Kuo (1031–1098) was a scientist and politician who lived during the Northern Song Dynasty. His famous work *Dream Stream Essays* (*Meng Xi Bi Tan*) covered a wide range of topics, including mathematics, astronomy, cartography, weather, geology, geography, biology, chemistry, water conservancy, agriculture, medicine, archaeology, linguistics, literature, music, painting and economics. It was an encyclopaedia of the science in ancient China, especially from the Northern Song Dynasty. The work was thus considered a milestone in the history of Chinese science.

Though it is said that the sun influences the rise and fall of the tides, this does not agree with my observations.
The movement of the tides correspond to the positions of the moon, hence I believe the tides are caused by the moon, not the sun.
In 1072, Shen Kuo was placed in charge of the meteorological service. He was also involved in setting the calendar for the central agency.
During his tenure, he improved the armillary sphere and water clock. He also drew up a new calendar called the Feng Yuan Calendar. In 1074, he was made the inspector for the western road in Hebei.
While inspecting the northern border, he made three-dimensional relief maps to help him report the terrain to the imperial court.
This melted wax is easy to model to show rivers and mountains. Now I only need bamboo sticks as markings on the board.
Shen Kuo also pioneered the use of a compass in map-making and listing straight-line distances instead of route lengths.

While in office, Shen Kuo designed a new suit of armour which arrows could not penetrate even from just 50 feet away.

In his twilight years, Shen Kuo drew up a atlas of the counties and provinces in China. It comprised 20 rolls and was the most complete map at that time.

He led a life of seclusion in the scenic Dream Stream Garden in Runzhou. He compiled his years of study and wide travels into a book called the *Dream Stream Essays*. Its wealth of content and scientific observations were way ahead of his contemporaries.

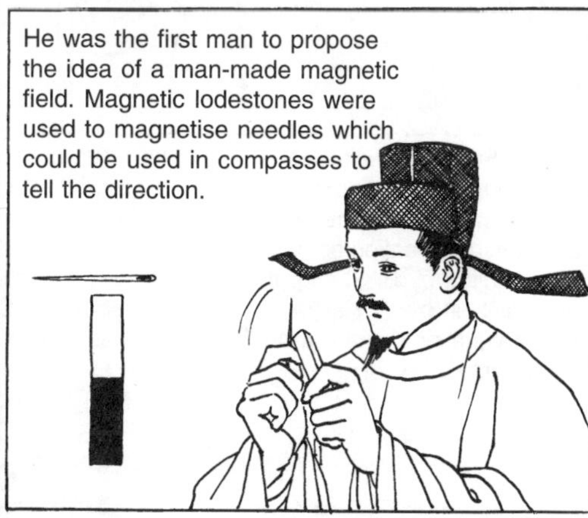
He was the first man to propose the idea of a man-made magnetic field. Magnetic lodestones were used to magnetise needles which could be used in compasses to tell the direction.

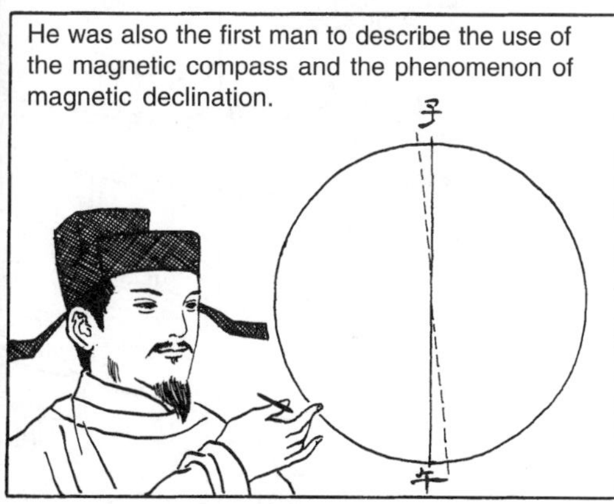
He was also the first man to describe the use of the magnetic compass and the phenomenon of magnetic declination.
子
午

Through his observation of shell, clam and sea urchin fossils on cliffs, he was able to infer the previous existence of an ocean which gradually became land over time. Europe came up with a similar theory only in the 18th century.

He divided the year into 12 portions and four seasons, and each season into three months — a division very similar to the modern calendar.

Dream Stream Essays also recorded the scientific achievements of the people, such as the invention of movable-type printing by Bi Sheng, the pagoda built by Yu Haojian, how floods were prevented through dams and so on.
十二氣
孟 仲 季

Shen Kuo made outstanding contributions in science, politics and military affairs. He was a blessing to the people. He relentlessly promoted water conservancy in agriculture, overhauled the salt industry and resolved the currency flow problem. As envoy to the Liao state, he resolved border conflicts and thwarted invasions from the state of Western Xia.

Guo Shoujing

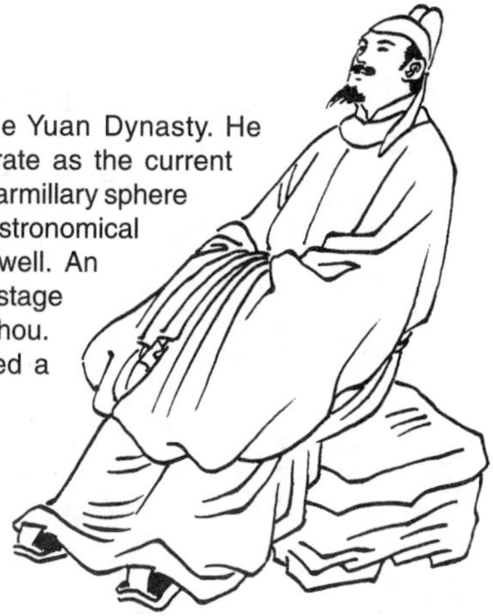

Guo Shoujing (123–1316) was a scientist from the Yuan Dynasty. He formulated the official calendar which was as accurate as the current Gregorian calendar. He improved on and simplified the armillary sphere invented during earlier dynasties. The famous Beijing Astronomical Observatory built in 1276 was designed by him as well. An expert in water conservancy, he completed the last stage of the Grand Canal running through Beijing and Hangzhou. In 1970, the International Astronomical Union named a crater on the moon after him.

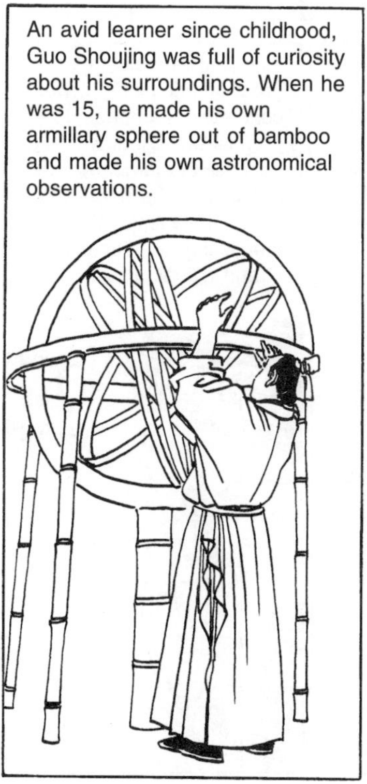

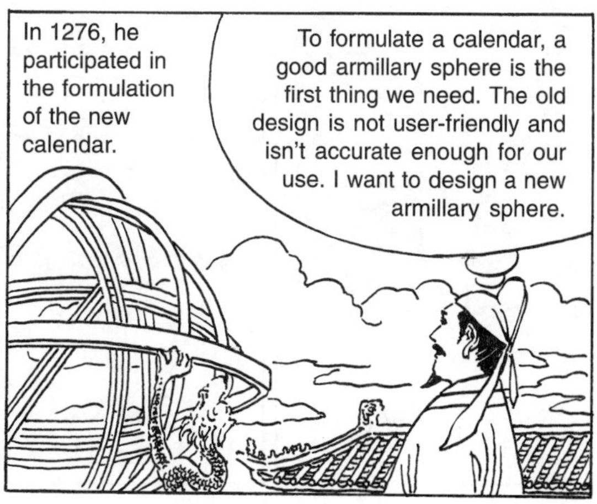

The old armillary sphere was a complex structure of many rings.
Remove these rings, but leave the two basic groups of rings intact. Dismantle the other rings to make independent structures.
Guo Shoujing's simplified armillary sphere was a milestone in the development of astronomical instruments. It was clean and simple. It was some 300 years before Danish astronomers came up with a comparable instrument.
Altogether, Guo Shoujing designed 20 astronomical instruments, allowing more accurate observations to be made and thus stamping his mark on Chinese astronomy.
He calculated that each year has 365.2425 days, a figure very close to today's Gregorian calendar, and redrew the calendar, naming it the *Official Calendar*. It was used for 363 years, longer than any other calendar used in China.
Guo Shoujing also made great contributions to water conservancy efforts. He proposed a canal which became the largest canal connecting the north and south of ancient China — the Bejing and Hangzhou Grand Canal.

Song Yingxing

Song Yingxing (1587–1661) lived during the Ming Dynasty. His most important work, *Exploitation of the Works of Nature* (*Tian Gong Kai Wu*), was a complete and systematic account of ancient Chinese agriculture, crafts and industry. The book contained many detailed illustrations and was dubbed the "encyclopaedia of Chinese skills and arts of the 17th century". It was regarded as an important text on the history of science and technology in the world, and later translated into Japanese, French and German.

Exploitation of the Works of Nature shows that all things are derived from nature. It is only when we recognise this that we are able to make use of nature to create goods and wealth.

... and observed the work procedures at the kiln.
It took 72 steps to make one small teacup.
He humbly enquired of every master artisan in every trade and industry.
Master, how did this piece of cloth get its colour?
Song Yingxing compiled his years of observations and study into the book *Exploitation of the Works of Nature*, which was published in 1637.
The book later reached Japan, Korea and countries in the west. European science and technology gained much technological know-how from it. For example, in the manufacture of brass, the wrong proportions of copper and zinc had rendered their previous attempts unsuccessfull. They learnt from the book that the ratio of copper to zinc should be 6:4 or 7:3. The technique of silkworm breeding also greatly benefited the western silk industry.
The book contained 18 volumes, categorised by topic. It was China's first book which combined technological knowledge and the skills of 18 industries, including agriculture and handicrafts, with a scientific approach.

Xu Guangqi

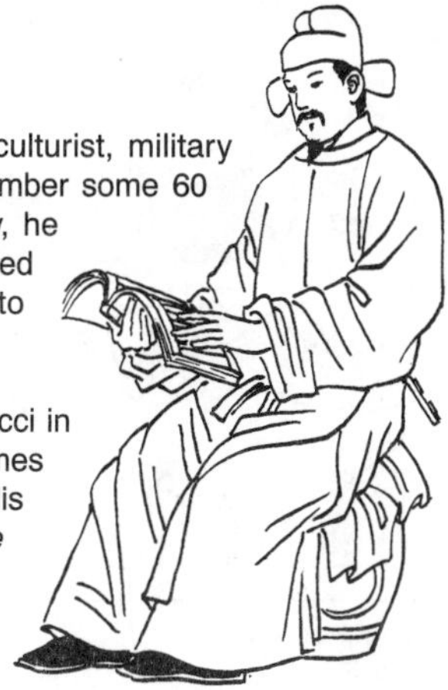

Xu Guangqi (1562–1633) was a scientist, translator, agriculturist, military expert and writer of the Ming Dynasty. In all, his works number some 60 books. Besides studying Chinese science and technology, he also absorbed scientific knowledge from beyond China and led the way in introducing western science and technology to China.

He collaborated with the Italian Jesuit missionary Matteo Ricci in translating several scientific texts, including the first six volumes of Euclid's famous mathematical treatise, *The Elements*. His own great work, *Nong Zheng Quan Shu* (*The Complete Agricultural Book*), was a landmark of China's ancient agricultural science.

I must get to know the person who drew the map and learn more about western science.
Xu Guangqi finally got his wish and met Ricci, who introduced him to Western science and Roman Catholicism. Xu Guangqi was converted.
Xu Guangqi often consulted Ricci on western science and technology.
What is this book?
This is *The Elements*. The author is Euclid, a Greek mathematician. It is a western mathematical classic.
This book explains geometric progression.
I tried to translate it into Chinese, but I just can't find the corresponding Chinese terms for the many terms and concepts involved.
What a pity! Why don't we try translating it together?
After working on it for a year, they finally translated the first six volumes and had them published. Xu Guangqi titled the book *The Original Manuscript for Geometric Progression*. It was the first translated book on western mathematics.

Many Chinese mathematical terms like *ji he, pingxing xian, duijiao xian, zhijiao* and *ruijiao* were coined by Xu Guangqi in the course of translation.
幾何原本
The Original Geometric Progression melded Chinese and Western concepts. It became a fundamental text for students of mathematics.
He was also very interested in agriculture.
The study of agriculture would benefit the people.
One year, a terrible flood occurred in Jiangnan and flooded the farms, destroying many crops. The people had no food to eat.
If a high-yield crop could be found, the people would not have to starve.
He heard that the Fujian sweet potato produced much higher yields than barley and decided to try growing it in Shanghai. After a few attempts, he succeeded.
Sweet potato is a high-yielding crop. It can be eaten raw or cooked. It can also be preserved by sun-drying. It's an ideal crop to alleviate famine.

He also wrote a book about the sweet potato.
甘薯疏
Xu Guangqi introduced the planting of turnips and improved cotton-growing techniques. He encouraged people to grow crops like the glossy privet tree and the Chinese tallow tree, which had greater economic value.
He also carried out agricultural experiments in Tianjin and introduced western hydraulic engineering methods. That was key to the development of Chinese agriculture.
He devoted his lifetime to the study of agriculture. Reading up on ancient agriculture and drawing on the experiences of farmers from various parts of the country, he collated the information and compiled it into a colossal work.
He also fine-tuned the calendar. Using the strong points of the western calendar to compensate for the shortcomings of the Chinese calendar, he took five years to formulate a new calendar called the *Chongzhen Calendar*.

The Scientist Emperor — Kangxi

Emperor Kangxi of the Qing Dynasty was an outstanding emperor. Not only was he an able ruler, he was also instrumental in stimulating many scientific achievements. He was particularly interested in mathematics and ordered that the textbook he used, *Shu Li Jing Yun* (*The Essence of Mathematics*) be translated from the Manchurian language into Han Chinese.

He set up a chemistry laboratory in the palace to concoct western medicines.

He also ordered a nation-wide cartography expedition, which resulted in *The Atlas of the Whole Imperial Territory*, also known as the *Kangxi Atlas*. It was used as the basis of maps by cartographers from other countries.

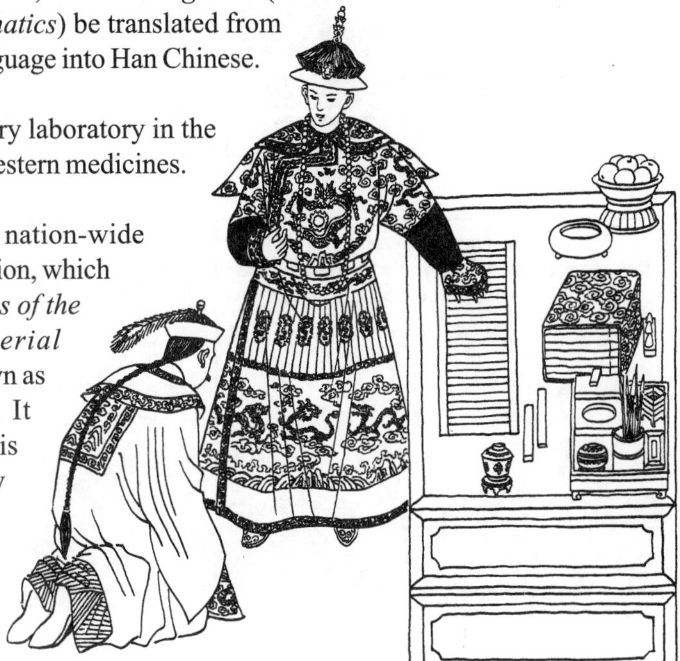

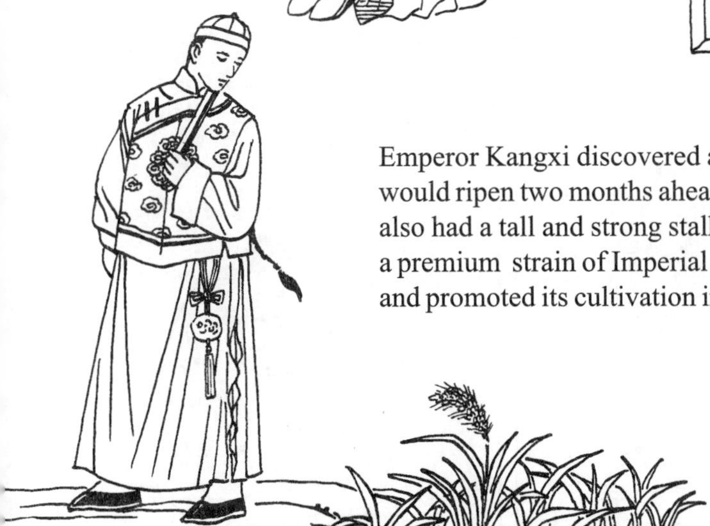

Emperor Kangxi discovered a particular grain that would ripen two months ahead of others. The grain also had a tall and strong stalk. He later developed a premium strain of Imperial Grain from this grain and promoted its cultivation in the south.

Inventors of Legend

Exchange of Science and Technology

In the course of cultural exchange with the outside world, the Chinese have incorporated foreign technology into their own and likewise, the latter has also been introduced to Chinese technology. The exchange of technological information has contributed to the advancement of world civilisation.

Introduction of Chinese Science and Technology to Japan and Korea

Given the geographical proximity of China, Japan and Korea, Chinese science and technology found their way to Japan and Korea quite early. Chinese copper smelting and metallurgical techniques arrived in Japan during the Western Han Period. Copper smelting and paint-making were two of many techniques introduced to Korea during the Eastern Han period.

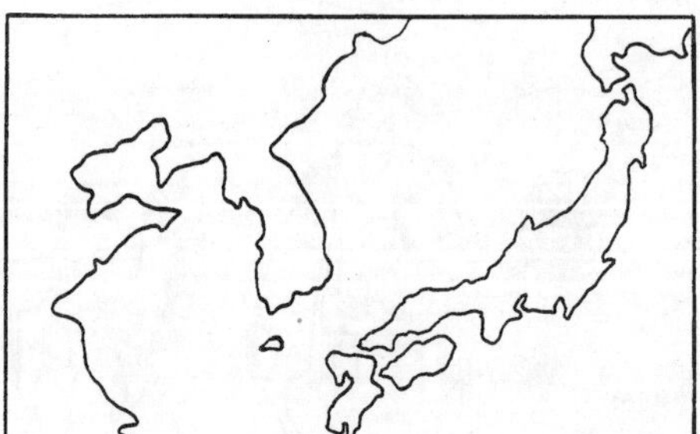

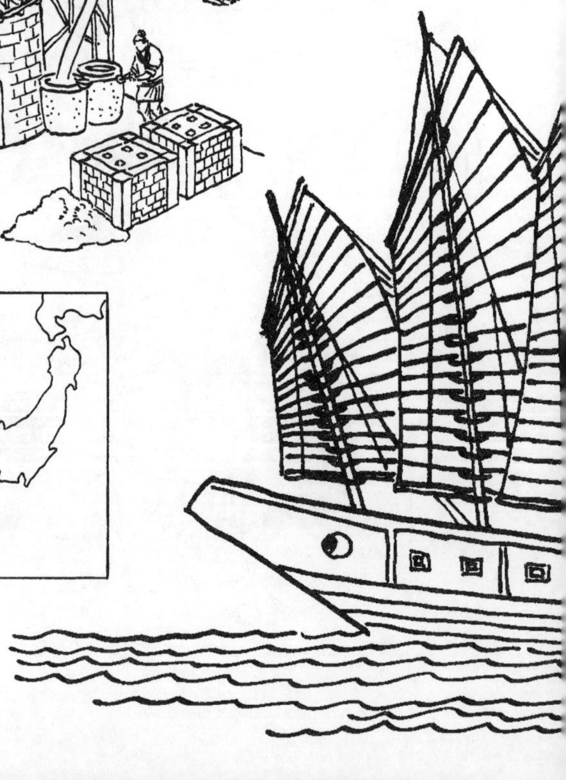

During the Sui and Tang Dynasties, China and Japan enjoyed close ties. Chinese carving, paper-making, printing, waterwheel-making, pottery and weaponry manufacture were among the technological skills introduced to Japan during the Tang Dynasty. Chinese astronomy, the calendar, mathematics and arts were also introduced to Shilla (then a kingdom in the Korean peninsular).

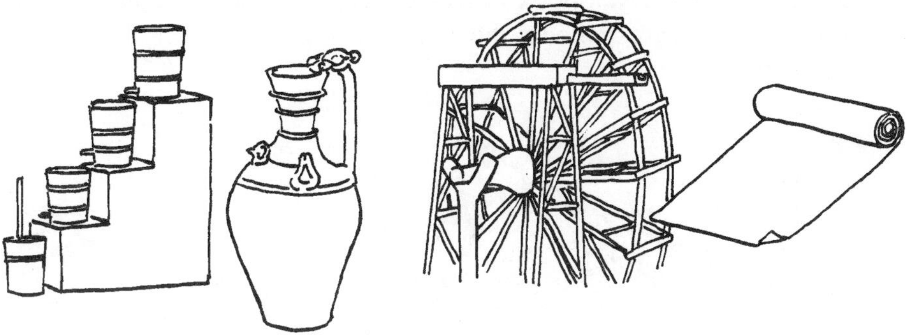

The Silk Road — Western Region

During the time of Emperor Hanwu, Zhang Qian was twice sent to the Western Region as an envoy. Along the way, he also visited Central Asia, West Asia and other regions. This was the beginning of the Silk Road, which opened the way for technological and cultural exchange between the Chinese and the Western Region during the Han Dynasty. Skills such as Chinese metallurgy, canal-building and well-digging were exported to the Western region, along with silk products and metalware.

The Story of the Silk Road
Head for the Western Region. I wish to form an alliance with Dayueshi against the Huns.
Zhang Qian left Xi'an with an entourage 100-strong. But the Huns captured and held them for 10 years.
In 119 BC, Zhang Qian made his second diplomatic trip to the Western Region. This time, he took more than 300 men with him and visited many places in the Western Region.
I have silk, satin, metalware and lacquerware from China. I've also brought various skills and techniques.
Your Excellency, we have obtained the seeds of grapes, pomegranates, walnuts and carrots.
They will be of great use to us.
The route he took, the Silk Road, was used by merchants, peasants, monks and other travellers moving between China and the Western Region.

Foreign Ties During the Sui-Tang Dynasties

The Sui and Tang Dynasties witnessed the golden age of Chinese culture and scientific developments. The governments of the dynasties encouraged an open door policy which boosted the scientific and cultural exchanges between the Chinese and the foreigners.

Paper-making and pill-making were introduced to the Indian peninsula during the Tang Dynasty, while the latter's knowledge of mathematics, astronomy and medical science were exported to China.

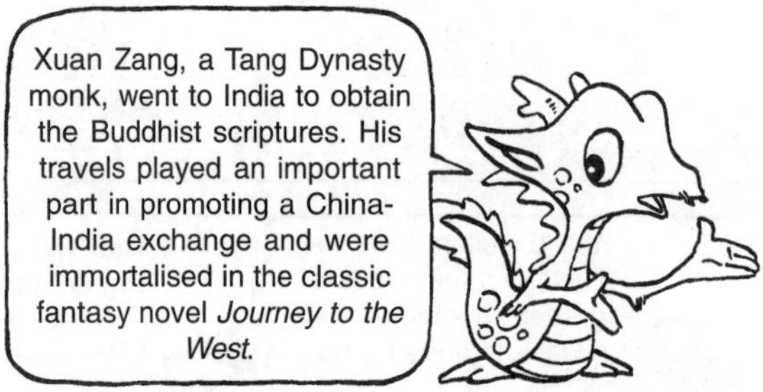

Chinese silk, pottery and paper were also introduced to Persia (present-day Iran) during the Sui and Tang Dynasties. In AD 751, war broke out between China and the Arabian countries in Central Asia. Many Tang soldiers who were also paper-makers and artisans with other skills were held as prisoners-of-war. They introduced the art of making paper and other skills to the Arabs. Chinese pottery-making and pill-making techniques were also introduced to Arabs during the Tang Dynasty.

During the Sui-Tang Dynasties, China also maintained ties with the Roman Empire. That was the time that Roman medical skills were introduced to China.

The Introduction of Western Science and Technology

The 16th and 17th centuries were revolutionary periods for western science. Breakthroughs were made in the study of astronomy, mathematics, biology, chemistry, medicine, natural sciences and etc.

From the late 16th century to the turn of the 18th century, western science and technology began to spread to China, mainly through the Jesuit missionary monks. In order to spread their faith, they used science and technology as their gateway into China.

The better known missionaries are Matteo Ricci (1552–1610) from Italy, the German astronomer Johann Adam Schall Von Bell (1591–1666), and the Belgian scientist Ferdinand Verbiest (1623–1688). While on their missionary quests, they also introduced modern western knowledge on astronomy, mathematics, geography, biology and firearms to the Chinese people.

Johann Adam Schall Von Bell
(Tang Ruo Wang)

Matteo Ricci
(Li Ma Tou)

Ferdinand Verbiest
(Nan Huai Ren)

The Future of Chinese Science and Technology

Throughout the history of China, there have been many times when conservative ruling powers closed the borders to foreign influences and discouraged scientific and intellectual endeavours. This caused China to fall back in the field of science and technology despite the many great innovations and discoveries of earlier times. Thus, the Chinese were amazed when western missionaries brought inventions such as the mechanical clock with them, forgetting they had also invented similar devices in the distant past.

The Qing Dynasty had promising beginnings, with Emperor Kangxi's interest in expanding the pool of knowledge, but during the 19th century, social and economic problems as well as corruption and conflicts with foreign powers led to a period of decline in technology and industry.

In more recent years, the Chinese have redoubled their efforts to modernise and catch up with the developments in other parts of the world. With the launch of China's first manned space mission in October 2003, the future of Chinese science and technology looks bright once more.

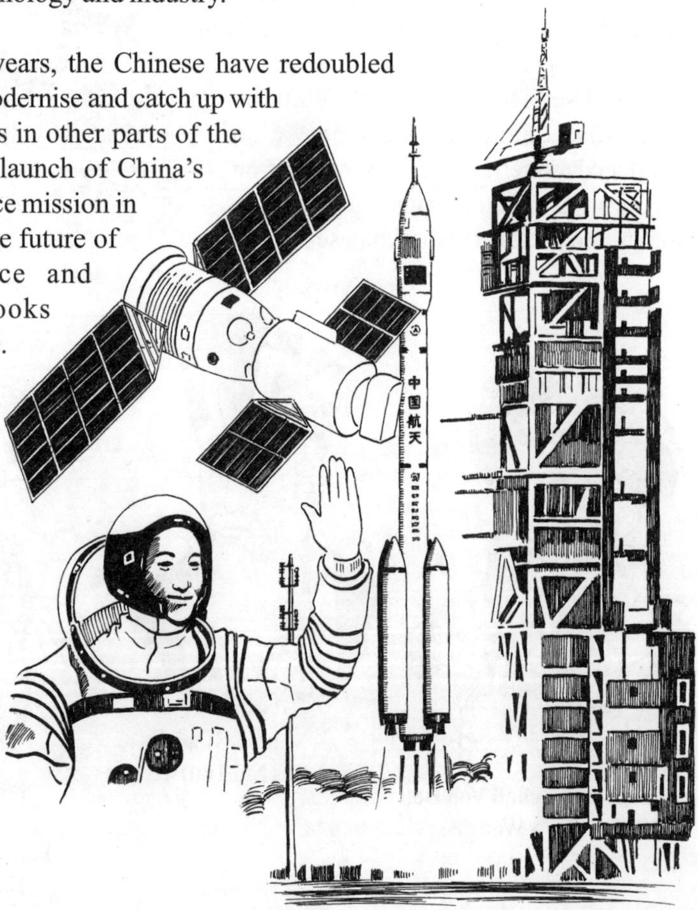

A Brief Chronology of Chinese History

夏 Xia Dynasty			About 2100 – 1600 BC
商 Shang Dynasty			About 1600 – 1100 BC
周 Zhou Dynasty	西周 Western Zhou Dynasty		About 1100 – 771 BC
	東周 Eastern Zhou Dynasty		770 – 256 BC
	春秋 Spring and Autumn Period		770 – 476 BC
	戰國 Warring States		475 – 221 BC
秦 Qin Dynasty			221 – 207 BC
漢 Han Dynasty	西漢 Western Han		206 BC – AD 24
	東漢 Eastern Han		25 – 220
三國 Three Kingdoms	魏 Wei		220 – 265
	蜀漢 Shu Han		221 – 263
	吳 Wu		222 – 280
西晋 Western Jin Dynasty			265 – 316
東晋 Eastern Jin Dynasty			317 – 420
南北朝 Northern and Southern Dynasties	南朝 Southern Dynasties	宋 Song	420 – 479
		齊 Qi	479 – 502
		梁 Liang	502 – 557
		陳 Chen	557 – 589
	北朝 Northern Dynasties	北魏 Northern Wei	386 – 534
		東魏 Eastern Wei	534 – 550
		北齊 Northern Qi	550 – 577
		西魏 Western Wei	535 – 556
		北周 Northern Zhou	557 – 581
隋 Sui Dynasty			581 – 618
唐 Tang Dynasty			618 – 907
五代 Five Dynasties	後梁 Later Liang		907 – 923
	後唐 Later Tang		923 – 936
	後晋 Later Jin		936 – 946
	後漢 Later Han		947 – 950
	後周 Later Zhou		951 – 960
宋 Song Dynasty	北宋 Northern Song Dynasty		960 – 1127
	南宋 Southern Song Dynasty		1127 – 1279
遼 Liao Dynasty			916 – 1125
金 Jin Dynasty			1115 – 1234
元 Yuan Dynasty			1271 – 1368
明 Ming Dynasty			1368 – 1644
清 Qing Dynasty			1644 – 1911
中華民國 Republic of China			1912 – 1949
中華人民共和國 People's Republic of China			1949 –

HISTORY

Imperial Chinese Women
The harem was a place where only the most beautiful, intelligent, fortunate and ruthless women could rise to the top. This book tells the stories of these women, the outstanding, the outrageous, the glorious and the tragic ones of the Chinese imperial harem.
160pp, ISBN 978-981-229-482-1

NEW

Chinese History: Ancient China to 1911
This book will help you to comprehend and interpret China's history in its proper context, plus provide vivid illustrations, and questions and answers to enhance your appreciation of great people and happenings.
192pp, ISBN 981-229-439-2

Great Chinese Emperors: Tales of Wise and Benevolent Rule
Read the tales of wise and benevolent rulers including Shennong, Li Shimin (Tang dynasty) and Emperor Kangxi (Qing dynasty). These stand tall for their outstanding contributions and character.
192pp, ISBN 981-229-451-1

Infamous Chinese Emperors: Tales of Tyranny and Misrule
Stories of China's most notorious emperors who are a motley crew of squanderers, murderers, thugs and lechers, and how they got their just deserts!
192pp, ISBN 981-229-459-7

PHILOSOPHY

ZEN INSPIRATION
Zen is a way of creative living. In this book, you will find out about Zen in all its vitality and simplicity. Whatever it is about Zen that fascinates you - silent meditation or creative expression - you will not be disappointed as you dip into the pages of this book.
Illustrated by **Fu Chunjiang**. *224pp, 150x210mm, ISBN 981-229-455-4.*

INSPIRATION FROM CONFUCIUS:
Choice Quotations from the Analects
More than 100 choice quotations classified under broad themes depicting Confucian core values and enhanced by inspirational thoughts. With additional features on Confucius' life, achievements and influence, it makes an excellent representation of the *Analects*.
Illustrated by **Jeffrey Seow**. *224pp, 150x210mm, ISBN 981-229-398-1.*

THE TAO INSPIRATION :
Essence of Lao Zi's Wisdom
Written more than 2,500 years ago, the Tao Te Ching now comes in 21st century style. Presenting Lao Zi's masterpiece in a concise, comprehensive yet profound manner, this book provides practical wisdom for leadership and for achieving balance and harmony in everyday life.
Illustrated by **Feng Ge**. *176pp, 150x210mm, ISBN 981-229-396-5.*

THE ART OF PEACE
The perfect companion if daily stories on war and terrorism are tiring you out. Learn how Mo Zi spread his message of peace to warring states locked in endless conflicts and power struggles.
Illustrated by **Chan Kok Sing**, *152pp, 150x210mm, ISBN 981-229-390-6.*

中华科技的故事

绘画　：傅春江

翻译　：杨立平、韩玉

亚太图书有限公司出版